THINKING GOD'S THOUGHTS

JOHANNES KEPLER
────── *and the* ──────
MIRACLE OF COSMIC COMPREHENSIBILITY

Portrait of Johannes Kepler

THINKING GOD'S THOUGHTS

JOHANNES KEPLER
and the
MIRACLE OF COSMIC COMPREHENSIBILITY

Melissa Cain Travis

Foreword by Stephen C. Meyer, PhD

Thinking God's Thoughts:
Johannes Kepler and the Miracle of Cosmic Comprehensibility

Copyright © 2022 Melissa Cain Travis

Published by Roman Roads Press
Moscow, Idaho

RomanRoadsPress.com
KeplerBook.com

General Editor: Daniel Foucachon
Editor: Carissa Hale
Interior Layout: Carissa Hale
Cover and interior illustrations: Joey Nance

Scripture quotations are from The ESV® Bible (The Holy Bible, English Standard Version®), copyright © 2001 by Crossway, a publishing ministry of Good News Publishers. Used by permission. All rights reserved.

All rights reserved. No part of this publication may be reproduced, stored in a retrieval system, or transmitted in any form by any means, electronic, mechanical, photocopy, recording, or otherwise, without prior permission of the publisher, except as provided by the USA copyright law.

For permissions or publishing inquiries, email info@romanroadspress.com

Thinking God's Thoughts: Johannes Kepler and the Miracle of Cosmic Comprehensibility
Melissa Cain Travis

ISBN 13: 978-1-944482-76-3
ISBN 10: 1-944482-76-8

Version 1.0.0 December 2022

Endorsements

"In a work of marvelous intellectual and historical retrieval, Dr. Travis has lucidly explained the great Kepler's significance for the philosophy of science and the philosophy of religion. Those interested in these disciplines as well as in the ongoing project of natural theology and Christian apologetics will be richly rewarded by *Thinking God's Thoughts.*"

Douglas Groothuis
Professor of Philosophy, Denver Seminary
Author of *Christian Apologetics: A Comprehensive Case for Biblical Faith*, 2nd ed. (IVP Academic, 2022)

"More than 400 years ago astronomer Johannes Kepler answered a question that still bothered Stephen Hawking: 'what is it that breathes fire into the equations and makes a universe for them to describe?' This new book by Melissa Cain Travis masterfully explores the development of these thoughts from ancient Greece to the fathers of modern quantum mechanics and shows how the natural philosophy of Johannes Kepler addressed those deep questions that are still confounding the greatest modern scientists from Einstein to Hawking: why is the cosmos comprehensible for us humans at all, and why can it be described so successfully with simple and elegant mathematical laws? The topic of *Thinking God's Thoughts* resonates deeply with the development of my own thought, as I was very much moved by the strange relationship between abstract mathematics and the physical world as well as the argument from reason. In my view these two issues represent the most powerful challenges to materialism and atheism. This beautiful book fills an important gap in the apologetic literature on arguments for the existence of God and is a must read for everyone interested in the great questions of life."

Dr. Günter Bechly, Ph.D.
Paleontology, Eberhard-Karls University of Tübingen
Senior Fellow at the Discovery Institute's Center for Science and Culture; scientific curator from 1999—2016 at the State Museum of Natural History in Stuttgart, Germany; senior scientist at Biologic Institute

We are so used to the fact that physicists use mathematics to both describe the world and to make discoveries at the subatomic and cosmological scales, that we take this mathematical intelligibility entirely for granted. This state of affairs is a great example of the phrase "familiarity breeds contempt." But as Melissa Cain Travis points out in this important book the user-friendliness of the universe screams for an explanation, and that cosmological comprehensibility is an enormous puzzle—unless the universe and its human inhabitants have been ultimately designed and created by a divine intelligence. Travis begins by showing how the idea of mathematical intelligibility is part of the Western Tradition's intellectual founding in the ideas of Pythagoras and the great Plato. She then describes in accessible detail how Kepler's life and work manifested this Pythagorean-Platonic worldview, and how Kepler used it to solve a millennia-old problem that was key to the Scientific Revolution and founding of modern science. But her book isn't merely a historical study; Travis also helpfully discusses the deep philosophical issues surrounding the mathematical intelligibility of the universe. But again, she does so in a way that is accessible to the non-specialist. Moreover, this book gives us a wonderful example of how STEM and the humanities are ultimately a single discipline, countering today's artificial siloing of academic subjects. I plan to use this book in my own courses.

Dr. Mitch Stokes, Ph.D.
Philosophy, University of Notre Dame

Senior Fellow of Philosophy at New Saint Andrews
College, Moscow, ID, and Head of Math, Science, & Music

CONTENTS

List of Photos	xiii
List of Figures	xiii
Foreword by Stephen C. Meyer	xv
Acknowledgements	xxiii
Introduction	1

PART I

The Development of Pythagorean-Platonic Philosophy Culminating in Johannes Kepler's Christian Natural Philosophy

Chapter 1: The Early Pythagorean Tradition	13
Pythagoras of Samos	14
Pythagoreanism According to Aristotle	23
Plato's *Timaeus* as a Key Transmitter of Pythagorean Thought	26
Chapter 2: Developments in Middle Platonism	33
Nicomachus of Gerasa	43
Chapter 3: The Pythagorean-Platonic Tradition from Ptolemy to Proclus	49
Ptolemaic Cosmology and Harmonics	50

The Pythagorean-Platonic Tradition in the Early Christian Era	55
Neo-Platonic Doctrine and Christian Thought	59
The Transmission of Neo-Platonic Thought Through Pagan Writings	67

Chapter 4: The Pythagorean-Platonic Tradition in the Middle Ages — 71

The Capstone: Nicolaus Copernicus	82

Chapter 5: The Early Formation of Kepler's Natural Philosophy — 89

Kepler's University Years	89
The *Mysterium Cosmographicum*	94
The *Astronomia Nova*	102

Chapter 6: Kepler's Magnum Opus—The *Harmonice Mundi* — 107

Excursus: Galileo and Newton	123

Chapter 7: The Scientific Fruitfulness of Kepler's Harmonies — 133

Kepler's Harmonization of Christian Theism and Natural Philosophy	139

Chapter 8: Kepler's Tripartite Harmony — 149

The Tripartite Harmony and Contemporary Natural Theology	159

PART II

Archetype, Copy, and Image: The Tripartite Harmony of Kepler's Natural Philosophy

Chapter 9: Archetype—God's Mathematical Plan for Creation   165
     Kepler's Understanding of the Archetype   166
     A Preliminary: Divine Aseity   168
     The Problem of God and Abstract Objects   170
     Theistic Platonism   173

Chapter 10: Three Main Alternatives to Theistic Platonism   181
     Option 1: Absolute Creationism   182
     Option 2: Theistic Conceptualism   187
     Option 3: Anti-realism   192
     Kepler's View   197

Chapter 11: Copy—The Mathematical Intelligibility of the Cosmos   199
     Philosophical Ponderings of Twentieth-Century Physicists   200
     Eugene Wigner and the "Unreasonable Effectiveness" of Mathematics   212

Chapter 12: The Scientific Applicability of Mathematics in Contemporary Thought   225
     An Anthropocentric Plan   225
     Penrose's Triangle   228

    Philosophical Implications of the Comprehensible
        Universe   235

Chapter 13: Image—Human Rationality and the Natural
    Sciences   241
    A Brief Intellectual History of the Argument from
        Reason   242
    C. S. Lewis's Formulation of the AR   253
    The Lewisian AR in Contemporary Philosophy   257

Chapter 14: The Success of the Natural Sciences as Evidence
    for Authentic Rationality   263
    Can Naturalistic Evolution Solve the Problem?   274
    Science, Therefore Incorporeal Minds   277

Conclusion   281

Appendix: Kepler's Remaining Years and Legacy   285
Endnotes   293
Bibliography   305
General Index   325

List of Photos

Portrait of Johannes Kepler	Frontispiece
A depiction of the Great Comet of 1577 over Prague	xxiv
Pythagoras (detail from School of Athens *by Raphael)*	15
The Kepler Monument in Regensburg, Germany Architect: Emanuel Herigoyen	291
Sketch of Kepler's Tombstone	302

List of Figures

Figure 1.1 *The Pythagorean Cosmology*	19
Figure 1.2 *The Tetractys*	25
Figure 1.3 *The Five Platonic Solids*	29
Figure 3.1 *The Aristotelian-Ptolemaic Cosmology*	51
Figure 4.1 *The Copernican Cosmology*	85
Figure 5.1 *Kepler's diagram of the Polyhedral Model from* Mysterium Cosmographicum	97
Figure 5.2 *Kepler's Three Planetary Laws*	106
Figure 6.1 *Kepler's 3rd Law Example*	112
Figure 6.2 *The Tychonic System*	125
Figure 12.1 *Penrose's Triangle—Three Worlds, Three Mysteries*	228

Foreword

by Stephen C. Meyer

"Well, of course you're an atheist, but what else are you that makes you interesting?" The question was posed to an acquaintance of mine by a faculty member at Cambridge University. At the time, we were sitting together in a pub, relaxing after an intensive seminar. My acquaintance was a fellow graduate student who had wanted to curry favor by boasting that he had thrown off his religious upbringing and now considered himself an atheist. But the faculty member at our table was unimpressed. Of course intelligent people are atheists. Why point out the obvious?

This exchange is anecdotal, but hardly isolated. In fact, it is illustrative of a wider mindset in the Western world. To be rational—and scientific—is to be atheistic or naturalistic. The ascendancy of atheism (or naturalism) has long since been decided. There is no need to restate the obvious.

This view, like all myths, has its own origin story. It arises from many sources, which collectively span an array of academic disciplines, historical contingencies, and human experiences. But perhaps most prominent among these is a certain narrative—a particular story—about the *history of science*. The story is one of emancipation, of breaking the ecclesial chains of dogma and superstition, so that an unencumbered study of the natural world may rise triumphant. John William Draper famously captured the spirit of this narrative in his 1874 book, *History of the Conflict Between Religion and Science*: "…Christianity and Science are recognized by their respective adherents as being absolutely incompatible; they cannot exist together; one must yield to the other; mankind must make its choice—it cannot have both."[1] Science runs contrary to biblical religion. Over time, atheism and naturalism become the default, the baseline commitment of learned, intelligent people.

So, is there anything else about you that is interesting?

The Conflict Myth Unmade

Yet this narrative is false. Melissa Cain Travis's *Thinking God's Thoughts* provides a clear, thorough, and thoughtful analysis of one of the leading pioneers of modern science, Johannes Kepler, and the deep role of faith and theology that undergirded his scientific research—and that persists to the present day. For Kepler, science did not point to atheism.

Kepler was not alone. Robert Boyle, Isaac Newton, Galileo Galilei, Nicolaus Copernicus, and many others who established modern science were deeply religious thinkers. They didn't see any conflict between science and religion. On the contrary, they thought that by doing science, they were discovering God's design and revealing it to humankind. Indeed, it's no exaggeration to say that the Judeo-Christian religious tradition led directly to modern science.

To back up this claim, Cambridge University historian of science Joseph Needham posed a famous "Why there? Why then?" question. Why there—in Europe? Why then—in the sixteenth and seventeenth centuries? Why didn't modern science start some-

---

1 John William Draper, *History of the Conflict Between Religion and Science* (New York: Appleton, 1874), 363.

where else before then? After all, the Egyptians erected pyramids. The Chinese invented the compass, block printing, and gunpowder.

Romans built marvelous roads and aqueducts. The Greeks had great philosophers. Yet none of these cultures developed the systematic methods for investigating nature that arose in Western Europe during the sixteenth and seventeenth centuries.

This realization led Needham and other historians of science to look for some other "X factor" to explain why "the scientific revolution" occurred where and when it did. Here is the conclusion they reached: only the Judeo-Christian West had the crucial ideas that enabled the invention of science. As historian Ian Barbour says, "science in its modern form" arose "in Western civilization alone, among all the cultures of the world," because only the Christian West had the necessary "intellectual presuppositions underlying the rise of science."[2]

So, what were those presuppositions? We can identify three. As Melissa Cain Travis shows, all have their place in Kepler's seminal works. More generally, all find their origin in the Judeo-Christian idea of a Creator God Who fashioned human beings and an orderly universe.

Intelligibility

First, the founders of modern science assumed the intelligibility of nature. They believed that nature had been designed by the mind of a rational God, the same God Who also made the rational minds of human beings. These thinkers assumed that if they used their minds to carefully study nature, they could understand the order and design that God had placed in the world. As the British

---

2 Ian Barbour, *Religion and Science: Historical and Contemporary Issues* (San Francisco, CA: HarperSanFrancisco, 1997), 27.

philosopher Alfred North Whitehead famously argued, "There can be no living science unless there is a widespread instinctive conviction in the existence of an *Order of Things*. And, in particular, of an *Order of Nature*."[3] Whitehead particularly attributed this conviction among the founders of modern science to the "medieval insistence upon the rationality of God."[4] It was God's own rationality that undergirded both the human intellect *and* an orderly world—and the deep affinity between them.

This meant that the world was intelligible. It was created for *human* discovery, and these discoveries pointed humans back to the Creator. As philosopher Holmes Rolston III states, "It was monotheism that launched the coming of physical science, for it premised an intelligible world, sacred but disenchanted, a world with a blueprint, which was therefore open to the searches of the scientists. The great pioneers in physics—Newton, Galileo, Kepler, Copernicus—devoutly believed themselves called to find evidences of God in the physical world."[5]

As Melissa Cain Travis amply documents, Kepler himself spoke in much the same way. For example, he wrote:

> For [God] Himself has let man take part in the knowledge of these things and thus not in a small measure has set up His image in man. Since He recognized as very good this image which He made, He will so much more readily recognize our efforts with the light of this image also to push into the light of knowledge the utilization of the numbers, weights, and sizes which He marked out at creation. For these secrets are...set out before our

---

   3   Alfred North Whitehead, *Science and the Modern World* (New York: Free Press, 1925), 3–4, original emphasis.

   4   Ibid., 12.

   5   Holmes Rolston III, *Science and Religion: A Critical Survey* (Philadelphia: Temple University Press, 1987), 39.

eyes like a mirror so that by examining them we observe to some extent the goodness and wisdom of the Creator.[6]

The Contingency of Nature

Second, early pioneers of science presupposed the contingency of nature. They believed that God had many choices about how to make an orderly world. Just as there are many ways to design a watch, there were many ways that God could have designed the universe. To discover *how* He did, scientists could not merely deduce the order of nature by assuming what seemed most logical to them; they couldn't simply use reason alone to draw conclusions, as some of the Greek philosophers had done. For example, the Greeks thought that since the most perfect form of motion was circular, they assumed that the planets must have circular orbits—something Kepler later refuted by calculations based on careful observations. As Melissa Cain Travis puts it, Kepler "emphasized... the primacy of empirical data over philosophical precommitments," including "the circular orbits of the Aristotelian-Ptolemaic model."

Instead, scientists like Kepler thought that the order in nature was the product of divine deliberation and choice, what the Scottish theologian Thomas Torrance calls "contingent rationality."[7] Indeed, because of this theological belief, these scientists assumed they would have to observe, test, and measure in order to understand God's design. As historian of science Ian Barbour succinctly

---

6 Johannes Kepler, *Epitome of Copernican Astronomy*, as cited in Max Caspar, *Kepler*, trans. C. Doris Hellman (New York: Dover, 1993), 381.

7 Thomas F. Torrance, *Divine and Contingent Order* (Oxford: Oxford University Press, 2000), 3.

notes, "The doctrine of creation implies that *the details of nature can be known only by observing them.*"[8]

Notably, Kepler championed this foundational element of modern science. As historian of science Michael Keas explains:

> Kepler proposed that among the mathematical ideas that exist in the divine mind, God freely selected some of them to govern his creation. Because God has the freedom to make many possible universes consistent with his eternal attributes, one cannot simply deduce from prior principles a single way that God must have created the world. Consequently, detecting likely truth in scientific ideas requires testing, sometimes with experiments, so as to reveal the virtues and vices of theories.[9]

Like other pioneers of modern science, Kepler's theology provided fertile soil for an empirically rich approach to nature.

The Fallibility of Human Reasoning

Third, early scientists accepted a biblical understanding of the power *and* limits of the human mind. Even as these scientists saw human reason as a gift of a rational God, they also recognized the fallibility of humans and, therefore, the fallibility of human ideas about nature. As Steve Fuller explains:

> Research in the history and philosophy of science suggests two biblical ideas as having been crucial to the rise of science, both of which can be attributed to the reading of Genesis provided by Augustine, an early church father, whose work became increasingly studied in the late Middle Ages and especially the Refor-

---

8   Barbour, *Religion and Science*, 28, original emphasis.

9   Michael Newton Keas, *Unbelievable: Seven Myths about the History and Future of Science and Religion* (Wilmington, DE: ISI Press, 2019), 163.

mation. Augustine captured the two ideas in two Latin coinages, which *prima facie* cut against each other: *imago dei* and *peccatum originis*. The former says that humans are unique as a species in our having been created in the image and likeness of God, while the latter says that all humans are born having inherited the legacy of Adam's error, "original sin."[10]

Such a nuanced view of human nature implied, on the one hand, that human beings could attain insight into the workings of the natural world, but that, on the other, they were vulnerable to self-deception, flights of fancy, and prematurely jumping to conclusions. This composite view of reason—one that affirmed both its capability and fallibility—inspired confidence that the design and order of nature could be understood if scientists carefully studied the natural world, but also engendered caution about trusting human intuition, conjectures, and hypotheses unless they were carefully tested by experiment and observation.[11]

In astronomy, for example, skeptical thinkers argued that humans could not know the true motions of heavenly bodies. At best, they could merely "save the phenomena"—invent fictional theories that were meant to capture the empirical data without ever discerning the true causes of things. Yet Kepler demurred. He regarded astronomy as a search for the *truth* about heavenly motion, a genuine "celestial physics." As Melissa Cain Travis observes, Kepler "was the first [natural philosopher] to develop a true *physica coelestis* and to search for a universal law to account for celestial mechanics." Yet Kepler also knew that such a goal required very careful thinking. Melissa Cain Travis explains that Kepler's pioneering work on epistemic virtues and theory choice reflects this nuanced, attentive care. Like others, Kepler knew that God's design of the

---

10 Steve Fuller, "Foreword," in *Theistic Evolution: A Scientific, Philosophical, and Theological Critique*, edited by J. P. Moreland, et al. (Wheaton, IL: Crossway, 2017), 30, original emphasis.

11 See Peter Harrison, *The Fall of Man and the Foundations of Science* (Cambridge: Cambridge University Press, 2007).

human mind allowed great discoveries, but only if the fallibility of human nature could be carefully constrained.

The *Telos* of Science

So, the Judeo-Christian worldview provided crucial foundations for the rise of modern science. And Kepler was in the middle of it. But while a biblical view may be the basis of science, just what was the end of it all? That is, what was the deeper purpose of the study of nature? As *Thinking God's Thoughts* demonstrates, the study of things below was originally meant to point humans to the splendor of things above. Moreover, Melissa Cain Travis's application of Keplerian insights to contemporary discussions shows that the true *telos* of science is *still* to point to God.

Perhaps Kepler said it best in his prayerful reflections about his astronomical demonstrations in *The Harmony of the World*: "O Thou who by the light of Nature movest in us the desire for the light of grace, so that by it thou mayest bring us over into the light of glory…be gracious and deign to bring about that these my demonstrations may be conducive to Thy glory and to the salvation of souls, and may in no way obstruct it."[12]

---

12 Johannes Kepler, *The Harmony of the World*, E. J. Aiton, A. M. Duncan, and J. V. Field, eds. (Philadelphia: American Philosophical Society, 1997), 491.

Acknowledgements

I would like to express special appreciation to the following scholars, all of whom offered valuable commentary and critique during my research and writing process: Dr. Mark Linville; Dr. Gary Hartenburg; Dr. Robert Woods; Dr. Paul Gould; and Dr. J. P. Moreland. In addition, I would like to acknowledge that Dr. John Lennox has greatly inspired my love of this subject matter and has significantly shaped my thinking about the harmony between science and Christianity. I am eternally grateful for his work. Finally, my heartfelt thanks to my husband, Jonathan, for his encouragement and unceasing, enthusiastic support.

A depiction of the Great Comet of 1577 over Prague.

Introduction

On a clear night in 1577, a small German boy was led up to a high place by his mother. From that vantage point, they beheld the wondrous sight of a great comet passing near the earth, its incandescent tail arcing far across an indigo sky. This magical experience left an imprint upon the boy's soul, and when he wrote about it retrospectively decades later, he must have been struck by the fact that one of the precious few happy memories of his childhood reflected an early intimation of his divine calling to natural philosophy.

Johannes Kepler was born—prematurely—on December 27, 1571, the annual feast day of St. John the Evangelist.[1] His early home life in Weil der Stadt was tumultuous, to say the least. His father, Heinrich Kepler, was a hot-blooded, bellicose young man who answered the siren call of battle in 1574, leaving behind his family to join the well-paying Spanish Catholic army.[2] The following year, Kepler's mother Katharina decided to follow her husband

---

1 Kepler's birth date seems almost providential, considering the themes of divine reason (*Logos*) in John's gospel and the astronomical phenomena described in his apocalypse. It is quite fitting that Kepler was St. John's namesake. There is some speculation that Kepler's mother was pregnant prior to her marriage, but it is clear that Kepler firmly believed he was merely born eight weeks early.

2 Ulinka Rublack, *The Astronomer & the Witch: Johannes Kepler's Fight for his Mother* (New York: Oxford University Press, 2015), 23. Rublack notes that it was not unusual for Lutherans to fight against Calvinists in those years, but that Heinrich's decision probably had nothing to do with religious convictions.

abroad, leaving Johannes and his baby brother behind with their paternal grandparents.[3] Before his parents' eventual return in 1576, little Johannes suffered a near-fatal case of smallpox, and was left with disfigurements and permanently damaged eyesight. Before the boy reached adulthood, several of his siblings died in infancy and his father (after multiple absences) vanished for good in 1589.

When Kepler became old enough to attend school, he was sent to one of the Duchy of Württemberg's many German schools, which offered basic education in the common language. However, his teachers soon recognized his remarkable intelligence and transferred him to the superior Latin school, where students quickly gained complete fluency and read Latin classics such as the comedies of Terence by year three.[4] Unfortunately, Kepler's school attendance was repeatedly interrupted by relocation and financial hardship, sometimes for months or even years at a time. Preeminent Kepler scholar Max Caspar offers a poignant picture of the circumstances: "The boy was used for hard agricultural labor and had to help himself during these interruptions as best he could."[5] Although it took him five years, Kepler completed the three classes of study at the Latin school, then passed the state examinations required for entrance into one of the seminaries. These were Lutheran institutions, set up in former monasteries, designed to prepare future theologians and clergymen for the academic rigor of the University of Tübingen.

At the age of thirteen, Kepler entered the lower seminary where his mastery of Latin was critical. Latin was both spoken and written, and at this stage students were also expected to achieve fluency in Greek. Over the next few years, which included a stint

---

3 Max Caspar mentions Johannes being left behind with his grandparents, but Rublack reports that it was a friend of Katharina who cared for Johannes and his younger brother. See Ulinka Rublack, *The Astronomer & the Witch*, 24.

4 Max Caspar, *Kepler*, (New York: Dover, 1993), 37.

5 Ibid.

at the most competitive boarding school in Württemberg, Kepler completed a course of study that included a progression through the traditional liberal arts curriculum. He was deeply immersed in both Scripture and the classics, notably Cicero and Virgil, and began composing original poetry with the hope of writing comedies. A youth of devout faith, he received religious instruction with great zeal, and was deeply grieved by the heated theological disunity in the church, between Lutherans and Calvinists as well as between Protestants and Catholics. He already possessed a spiritually tender and ecumenical heart, which would ultimately affect the trajectory of his life.

Kepler passed the university entrance examination in 1588 and received one of five ducal scholarships to attend the *Stift*, the University of Tübingen's seminary, where he matriculated in the autumn of 1589. Details of the curriculum will be discussed in a later chapter as part of an in-depth discussion of the philosophical, literary, and theological traditions that shaped his intellect and ignited his imagination. Here, it need only be mentioned that Kepler demonstrated incredible academic agility; in addition to his grasp of mathematics and astronomy, he developed an outstanding command of the Greek language and showed promise as a rhetorician and philosopher. He continued to nurture his love of poetry, publishing verses and presenting printed copies of imaginative poems to his friends to mark special occasions.[6] During these years he formed important relationships with some of his esteemed professors, including a young theology professor by the name of Matthias Hafenreffer, and Magister Michael Maestlin, whose astronomy lectures contributed to Kepler's conclusion that the Copernican model of the universe was mathematically superior to the long-reigning Ptolemaic (geocentric) scheme. Kepler and Maestlin maintained a profitable professional correspondence for many years; the teach-

---

6 Ibid., 45.

er often advised and advocated for his erstwhile pupil, helping to open scholarly doors that may have otherwise remained closed. Kepler and Hafenreffer (whom Kepler believed to be a secret Copernican) remained close friends until the latter's death.

Kepler was awarded the *Magister Artium* in August of 1591 but remained at the *Stift* to further his theological studies, which were unexpectedly interrupted by an offer for a professional post as a mathematics instructor at a Protestant school in Graz. Making this decision was agonizing; for more than a decade he had worked diligently to equip himself for a position in the church, and he was approaching completion. Moreover, he was passionate about the study of sacred theology and was acutely aware of the comparative modesty of the prospective teaching post. On the other hand, he was cognizant of his mathematical talent and did not wish to exhibit the trepidation of fellow students who had made all sorts of excuses to avoid leaving their homeland: "I, being hardier, quite maturely agreed with myself that whithersoever I was destined I would promptly go."[7] So off to Graz he went, using borrowed money for travel expenses, to teach mathematics to teenage sons of Protestant aristocrats.

Such were the humble beginnings of Johannes Kepler's professional life. Already, he was on a far different course than the one he had expected, but his excellent liberal arts education, poetic soul, keen intellect, and yearning for knowledge of God would still prove essential, as will be seen. His path to becoming a giant of the scientific revolution would be marked by moments of intellectual and spiritual euphoria (for Kepler, these were often one and the same), heartbreaking tragedy, political upheavals, and maddening injustices. More of his personal story will unfold in later chapters and conclude in the biographical appendix. Presently, we turn to the main objective of this book.

---

[7] Johannes Kepler, *Astronomia Nova*, trans. William Donahue (Santa Fe, NM: Green Lion Press, 2015), 133.

A multitude of philosophers, theologians, scientists, and mathematicians of the great Western Tradition have been struck by the uncanny interconnection between three fundamentally distinct domains of reality: nature, mathematics, and the human mind. This resonance—which is responsible for cosmic comprehensibility—has been discussed since antiquity and often attributed to a transcendent rational source of both material and immaterial aspects of reality. Important roots of this idea can be traced back to the Pythagorean school and then forward, through Plato and the subsequent writings of Middle Platonists and neo-Platonists[8], which included pagan, Jewish, and Christian thinkers. This intellectual thread is part of the broader Pythagorean-Platonic tradition, and it reached a crucial culmination in the work of Kepler.

Kepler was instrumental in transforming classical astronomy, which was based upon Aristotelian philosophy and the Ptolemaic geometrical model, into a true celestial physics. Along the way, he reinforced traditional notions about the order and harmony of the universe and our extraordinary place in it. He perceived a three-part cosmic harmony of *archetype*, *copy*, and *image* that enabled mankind to unlock the secrets of nature.[9] The mathematically rational plan for creation—the archetype—existed from eternity in the mind of God; the material manifestation of the archetype is the copy that natural philosophers investigate, and the image of God in human beings includes the higher rationality that enables us to study nature using observation, mathematics, and innate preferences for beauty and simplicity. This tripartite harmony was Kepler's decidedly theistic explanation for the marvelous resonance between the rational

---

8 Some writers use Neoplatonist/Neoplatonism, while others use neo-Platonist/neo-Platonism. For the sake of stylistic consistency, the latter terminology is employed throughout this book.

9 The English translation of Max Caspar's seminal work on Kepler uses the term *prototype* rather than *archetype*, but following most English translations of Kepler's writings and related commentary, *archetype* will be used.

order of the material world, mathematics, and the human mind, and granted the natural philosopher the ability to share in God's own thoughts (as he famously put it). In the centuries since Kepler, the resonance necessary for cosmic comprehensibility has continued to inspire a high degree of wonder, even in some who reject theism. We see evidence of this in the writings of great thinkers of the later Western Tradition as well as the philosophical reflections of a diversity of contemporary scientists and philosophers.

Given Kepler's revolutionary discoveries in natural philosophy paired with his deep conviction that theism accounted for the very existence of his discipline, it seems appropriate to refer to this explanation for cosmic comprehensibility as *Keplerian natural theology*. The term *natural theology* has been used in various ways, so it is necessary to offer clarification here at the outset. Generally speaking, it is an inquiry into what can be known about God apart from what He has revealed about Himself in Scripture. As philosopher William Alston puts it, natural theology is "the enterprise of providing support for religious beliefs by starting from premises that neither are nor presuppose any religious beliefs."[10] These premises may include observations about the natural world, such as those employed in classic and contemporary design arguments. Keplerian natural theology is specifically concerned with two features of the world that are essential to the existence of the natural sciences: the mathematical structure of the material realm and the human intellectual capacity for advanced mathematical reasoning. As will be seen, it may be argued that the best explanation for the mathematics-nature-mind resonance is the existence of a Creator in Whose image we are made.

This book has three main goals. First, to outline the rich intellectual heritage that influenced Keplerian natural theology by examining the Pythagorean-Platonic themes that Kepler absorbed

---

10 William P. Alston, *Perceiving God: The Epistemology of Religious Experience* (Ithaca, NY: Cornell University Press, 1991), 289.

during his education; second, to explore his achievements in natural philosophy and the guiding role played by his natural theology; and third, to demonstrate that the core ideas of Keplerian natural theology have in no way been undermined by subsequent scientific advancement. Quite the contrary! In light of discoveries in fields such as cosmology and theoretical physics, as well as developments in certain areas of philosophy, Keplerian natural theology is a more robust explanation for cosmic comprehensibility than ever before; it brings about a superior coherence and alleviates several serious philosophical problems, including the highly ordered fundamental structure of nature and mankind's advanced aptitude for applicable mathematical reasoning. By contrast, naturalistic explanations have proven to be gravely insufficient.

Keplerian natural theology should be recognized as one of the so-called *great ideas* of the Western Tradition, since it boasts both an outstanding intellectual pedigree and undiminished vitality in the contemporary conversation. The *Keplerian* qualifier seems more than justified; Kepler's work in natural philosophy was pivotal in transforming ancient astronomy into a true mathematical physics of astronomy, and his entire life's work was driven by his belief in the fundamental harmony of archetype, copy, and image. In the most important sense of the word, Kepler *was* a theologian; he sought to understand the Creator through nature and believed that the natural philosopher is a priest in God's cosmic temple, worshiping and glorifying God by exegeting the book of nature. His deepest conviction was that his work in astronomy was a kind of communion that yielded dazzling glimpses into the very mind of God.

In recent years, the mysterious correspondence between nature, mathematics, and mind has been highlighted by thinkers of various religious persuasions, but there has not yet been a focused effort to weave these ideas together into one coherent, interdisciplinary project. Kepler is rarely (and then only briefly) mentioned in current literature on cosmic comprehensibility, and the intellectual tradition that culminated in his work has not received the attention it merits. Considering the ongoing and often heated debate about the compatibility of the natural sciences with Christian claims about the existence of God and mankind's privileged place in the world, it should be emphasized that the scientific enterprise would have been impossible in the absence of the mathematics-nature-mind resonance that delighted Kepler. From the existence of this astonishing state of affairs, it is more than reasonable to infer that there is, behind the world, a transcendent mind with whom we have a special kinship, who has revealed himself through the things he has made.

This book is divided into two main parts. Part I, which is comprised of eight chapters, explores the origin and development of the Pythagorean-Platonic tradition, which heavily influenced Kepler's astronomy and natural theology. It will begin with pre-Socratic Pythagoreanism and its later treatment in Plato's work, and then trace the development of the Pythagorean-Platonic tradition up through the philosophical movement known as Middle Platonism. This will be followed by a survey of philosophers and theologians who interacted with this tradition and whose work influenced Kepler directly or nearly so, beginning with Claudius Ptolemy and ending with Nicolaus Copernicus. Next, the intellectual milieu Kepler experienced at the University of Tübingen will be examined prior to a philosophical evaluation of several of his major works and his personal correspondence. Part I will culminate with an in-depth analysis of Kepler's natural theology, which was deeply integrated with his understanding of the material cosmos.

With this groundwork established, Part II will consider the ongoing vitality of Keplerian natural theology in the context of modern philosophical thought. First, the concept of *archetype*—the rational plan for creation pre-existing in the mind of God—will be evaluated in light of the contemporary debate about the problem of God and abstract objects. Next, the understanding of the natural world as a *copy* of the divine archetype will be examined within the context of the ongoing conversation regarding the mathematical comprehensibility of the universe. The third string of Keplerian natural theology, *image*, which says that mankind possesses the capacity for natural philosophy because of an intellectual kinship with the Creator, will be discussed in relation to current philosophical views on the higher rationality of man, giving particular attention to the plausibility of naturalistic accounts of mind. Part II will close with a few final remarks about why Keplerian natural theology remains a philosophically, scientifically, and theologically viable explanation for the grand cosmic resonance of nature, mathematics, and human rationality.

Part I

*The Development of Pythagorean-Platonic Philosophy
Culminating in Johannes Kepler's Christian Natural Philosophy*

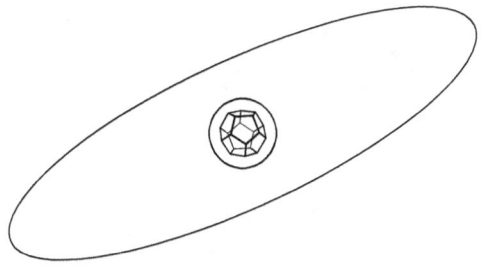

CHAPTER 1

The Early Pythagorean Tradition

Scholars and students of the Western Tradition have a keen interest in the emergence and development of the great ideas that, as Mortimer Adler once put it, "are basic and indispensable to understanding ourselves, our society, and the world in which we live."[1] Tracing thematic threads back to philosophers of deep antiquity is a fascinating endeavor. It often yields satisfying—and sometimes surprising—insights about how various ideas have persisted and morphed over time and how they have shaped crucial aspects of human life in successive historical periods. From time to time, revolutionary leaps in our understanding have occurred, but these leaps are not spontaneously generated *ex nihilo*; major intellectual advancements are progeny with long, complex pedigrees. One would be hard-pressed to find a better exemplar of this fact than the life's work of Johannes Kepler.

Max Caspar, who read and translated (Latin to German) every surviving word that flowed from Kepler's pen, regarded the astronomer as one "anointed with Pythagorean oil," a thinker whose natural philosophy was the grand culmination of what is referred to in scholarly circles as the Pythagorean-Platonic tradition.[2] More

---

1 Mortimer Adler, *The Great Ideas: From the Great Books of Western Civilization* (Chicago: Open Court, 2001), xxiii.

2 Max Caspar, *Kepler*, 380.

recently, Bruce Stephenson called Kepler's work "the last and most extravagant development of the Pythagorean theme,"[3] and historian of science J. V. Field described Kepler as the "last exponent of a form of mathematical cosmology that can be traced back to the shadowy figure of Pythagoras."[4] Thus, in outlining a selective intellectual history meant to show how Kepler's philosophical and theological ideas are situated within the Western Tradition, Pythagoras of Samos (fl. sixth century BC), a pre-Socratic who lived during the early infancy of Greek philosophy, seems to be an appropriate starting point.

Pythagoras of Samos

As explained in the introduction, Kepler's natural philosophy was guided by three core tenets. He believed that a mathematical plan—an archetype—for the created world pre-existed in the mind of God; that the material creation based upon this incorporeal model exhibits geometric and harmonic patterns that can be elucidated using the tools of mathematics; and that the constitution of the human mind is such that it resonates with the fundamentally mathematical structure of the cosmos. In other words, human beings possess precisely the type of cognitive faculties necessary for deciphering the abstract rational patterns and harmonies exhibited in nature. For Kepler, natural philosophy could be carried out only because of the rational order of the natural world paired with the fact that mankind is made in the image of God. He believed that the human mind is, in some finite sense, analogous with God's mind, and this allows the manifestations of the divine intellect in

---

3   Bruce Stephenson, *The Music of the Heavens: Kepler's Harmonic Astronomy* (Princeton: Princeton University Press, 1994), 11.

4   J. V. Field, *Kepler's Geometrical Cosmology* (New York: Bloomsbury, 1988), 170.

the material realm to be apprehended. The main objective of this chapter is to demonstrate that these three interlocking ideas are deeply rooted in the Western Tradition; they have discernible precursors in early Pythagorean thought, and their development can be traced forward through pre-Christian antiquity.

Although Pythagoras is the most famous philosopher of the pre-Socratic era, he is shrouded in tantalizing mystery. He left behind no writings, and some of the earliest references to him are ambiguous, satirical, or colored by the personal agenda of the doxographer. In the centuries following his death, a broad body of pseudepigraphal works and largely fictitious biographies were written and disseminated, which compounded the problem of the historical Pythagoras. Consequently, little can be known about his life and his oral teachings with much certainty, and even defining the term "Pythagorean" in a pre-Platonic sense is notoriously difficult. There is a broad yet fairly polarized spectrum of scholarly opinion about Pythagoras the philosopher. Some regard him as a brilliant mathematician and proto-scientist who actually coined the term *philosophy*, identified the five regular polyhedra, and discovered the numerical ratios underlying musical harmonies.

Pythagoras (detail from School of Athens by Raphael)

Others have characterized him as a highly influential shamanistic figure of the Orphic variety who endorsed an odd, cultic lifestyle[5] and did not contribute to the mathematical arts in any meaningful way.[6] This debate is largely due to the fact that early on (around the mid-fifth century BC), Pythagoreanism splintered into many different schools scattered around the Greek world. This fragmentation meant that, despite having a common root, the philosophy of each sect would develop in unique ways over time.

Since the original words of Pythagoras are lost to history, there will never be a perfect consensus about the content of his teachings.[7] Fortunately, prominent scholars of pre-Socratic philosophy agree that significantly more can be known about Pythagoreanism as Plato and his original Academy encountered it. Of the earliest textual records of Pythagoreanism, a few surviving fragments attributed to Socrates's contemporary Philolaus (c. 470–385 BC) and Philolaus's student Archytas of Tarentum (c. 428–350 BC) are the only ones regarded as authentic.[8] In what follows, each will be briefly examined with a specific focus upon the ideas that went on to strongly influence Kepler's work. Then, relevant passages from Plato and Aristotle will be analyzed for further illumination of fourth-century BC Pythagorean thought.

---

5 For example, legend has it that Pythagoreans wore only white clothing and were prohibited from eating beans. Explanations for the dietary restriction vary, some more comical than others.

6 Charles Kahn, *Pythagoras and the Pythagoreans: A Brief History* (Indianapolis: Hackett Publishing, 2001), 1–3.

7 This is not to say that progress hasn't been made (see Endnote 1). A detailed discussion of this situation can be found in W. K. C. Guthrie, *A History of Greek Philosophy: The Earlier Presocratics and the Pythagoreans* (Cambridge: Cambridge University Press, 1988), 180–181.

8 Walter Burkert, *Lore and Science in Ancient Pythagoreanism* (Cambridge: Harvard University Press, 1972), 220. Carl Huffman's magisterial work, *Philolaus of Croton: Pythagorean and Presocratic* (New York: Cambridge University Press, 1993) confirms and extends Burkert's argument for fragment authenticity.

Philolaus was a philosopher of southern Italy—from either Croton or Tarentum—and the most direct evidence for dating his life comes from Plato's *Phaedo* dialogue. In section 61, Cebes asks, "Then tell me Socrates, why is suicide held to be unlawful as I have certainly heard Philolaus...affirm when he was staying with us at Thebes?"[9] This short passage indicates that Philolaus was a contemporary of Socrates, and since the dialogue itself recounts details about the day Socrates was executed in 399 BC and says nothing about the death of Philolaus, it is reasonable to assume that the latter was still living when Plato authored the dialogue. This places Philolaus approximately a century after Pythagoras, yet he was probably the first to write a book on Pythagorean philosophy.[10] Of the surviving fragments of *On Nature*, about half are controversial, but even conservative scholars affirm the authenticity of the remainder, some of which touch on ideas that are pertinent to this discussion.[11]

Fragment 1 of Philolaus, which is preserved in the *Lives* of Diogenes Laertius, reads: "Nature in the world-order (*kosmos*) was fitted together harmoniously (*harmochthê*) from unlimited things (*apeira*) and also from limiting ones (*perainonta*), both the world-order as a whole and all things within it."[12] Philolaus explains that the so-called unlimited and limiters—the fundamental principles (*archai*) of the world—are harmonized into a *kosmos* (an orderly whole). Fragment 4 indicates that this harmony is essentially numerical, and it is this quality that makes things knowable: "And indeed all things that are known have number. For it is not possible that anything whatsoever be understood or known without

---

9   Plato, *Phaedo*, 61e.

10   Evidence for the existence of this book is provided by a papyrus discovered in 1893 that contains excerpts from a book by Aristotle's pupil, Menon. Menon refers to a book by Philolaus that existed at that time. See Guthrie, *A History of Greek Philosophy*, 278 n. 3.

11   See Endnote 2.

12   Charles Kahn, *Pythagoras and the Pythagoreans*, 24.

this."[13] It must be asked: what exactly are *the unlimited* and *limiters* that are arranged in numerical harmony? It has been convincingly argued that *the unlimited* is, for Philolaus, the eternal primordial substance, and *limiters* are numbers and mathematical functions. In this view, the limiters are the boundaries imposed upon the unlimited.[14] Thus, the limiters act upon the unlimited in a way that produces a harmonious fitting together, an arrangement that makes them knowable.[15] This connection between the mathematical ordering of things and intelligibility should be carefully noted; it is a key idea that will be revisited many times henceforth.

Philolaus's cosmological system is described in Fragment 7, which has been preserved by Johannes Stobaeus (fl. c. AD 500), a Greek writer whose *Anthology* contains numerous fragments from earlier antiquity. Philolaus envisioned a central fire or "hearth" (*hestia*) around which all celestial bodies revolved in concentric orbits. First was a so-called "counter-earth" (*antichthōn*), a planet that remained geometrically opposite of the earth (which occupied the subsequent orbit) as it traveled in its circular journey around the central fire. After the earth came the moon, sun, Mercury, Venus, Mars, Jupiter, Saturn, and the outer sphere of fixed stars. This system attempted to "save the phenomena"—to harmonize observations with the philosophical assumptions of that time. It represents the first known instance of Greek cosmology that incorporated the five visible planets, and it was even used to explain some lunar eclipses.[16] It seems fair, then, to regard Philolaic cosmology as a

---

13 Carl Huffman, *Philolaus of Croton*, 172.

14 Daniel W. Graham, "Philolaus," in *A History of Pythagoreanism*, ed. Carl Huffman (Cambridge: Cambridge University Press, 2014), 53.

15 See Endnote 3 for further discussion on the role of limiters in Philolaus's cosmogony.

16 In a fascinating scholarly article, Daniel Graham has argued that Philolaus should be considered a true empirical astronomer because the postulation of an unobservable counter-earth can indeed (Graham claims) account for some lunar eclipses in such a system. See Daniel W. Graham, "On Philolaus's astronomy," *Archive for History of Exact Sciences* 69 (2015): 217–230.

PYTHAGOREAN COSMOLOGY

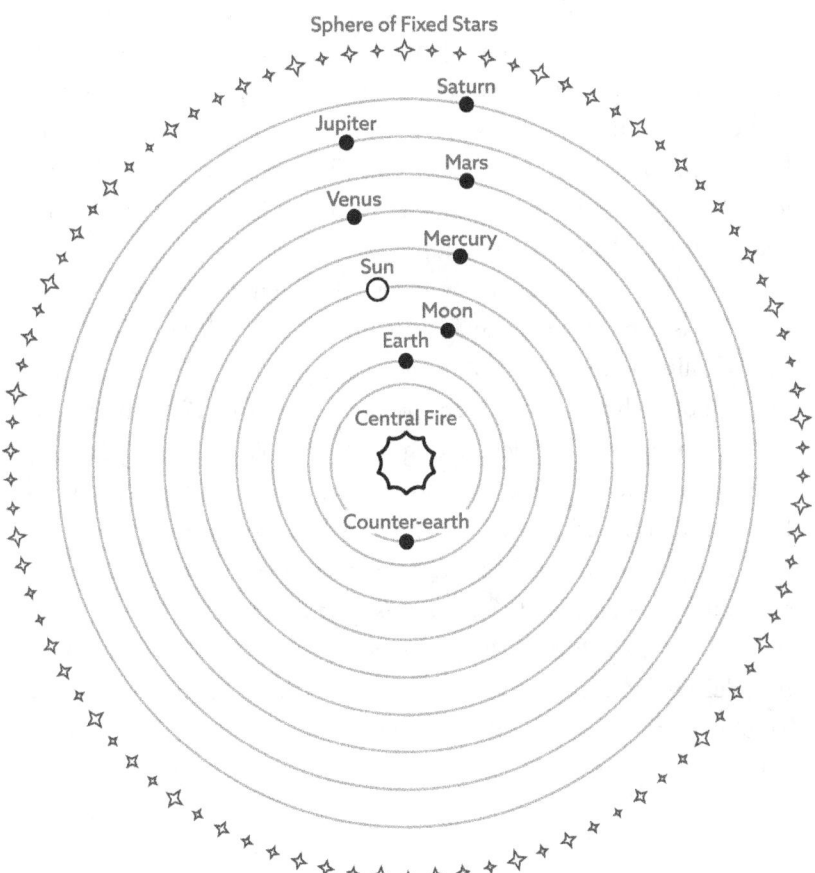

Figure 1.1 *The Pythagorean Cosmology*

very early scientific work (to use modern terminology) rather than an exercise in Pythagorean mysticism.

Philolaus's pupil, Archytas of Tarentum, is a crucial personal link between Philolaic Pythagoreanism and Plato's Academy. Archytas, who was known by Plato, was a highly successful statesman and undefeated military general who held the supreme office of *stratēgos* seven times, despite a law prohibiting more than one annual term. He was renowned for his military prowess, mild temperament, and excellent politics; it has even been suggested that Archytas may have been the real-life figure who inspired the portrait of the Philosopher King in Plato's *Republic*.[17] Tradition has it that around the age of forty, Plato first encountered Archytas on one of several journeys west to various Italian cities.[18] In the highly debated *Seventh Letter*, it is reported that in Plato's final visit (c. 361), he incurred the hostility of the tyrant of Syracuse, Dionysius II, and had to be rescued by a ship sent by Archytas and others from Tarentum.[19]

Archytas is regarded, along with Philolaus and Pythagoras himself, as one of the three most important philosophers of the Pythagorean tradition. Carl Huffman, whose monograph is the definitive work on Archytas, writes, "Archytas is a crucial figure for any attempt to understand Greek philosophy and mathematics outside of the Academy during this period and thus for understanding the broader environment in which both Plato and Aristotle developed as philosophers."[20] Huffman emphasizes Archytas's prominence within the higher intellectual milieu of the fourth century BC:

---

17 Guthrie, *A History of Greek Philosophy*, 333. Also see Carl Huffman, *Archytas of Tarentum: Pythagorean, Philosopher, and Mathematician King* (New York: Cambridge University Press, 2005), 44.

18 Huffman, *Archytas of Tarentum*, 33.

19 Ibid., 3. This account also appears in Diogenes Laertius's (third century AD) biography of Archytas as well as the tenth-century Byzantine encyclopedia known as the *Suda*.

20 Huffman, *Archytas of Tarentum*, xi.

Aristotle wrote more books on Archytas than any other individual figure. He devoted three books to the philosophy of Archytas himself and wrote another consisting of a summary of Plato's *Timaeus* and the writings of Archytas...Aristotle's pupil, Aristoxenus, appears to have begun the tradition of peripatetic biography and wrote a life of Archytas, thus putting him in the select company of Pythagoras, Socrates and Plato. Aristoxenus was from Tarentum and began his philosophical career as a Pythagorean, so it is not a surprise that he should choose to write a life of his countryman, but that choice also reflects the prominence of Archytas. Aristoxenus' contemporary, Eudemus, another pupil of Aristotle, referred to Archytas prominently in his history of geometry...and his physics.[21]

Unfortunately, Aristotle's books on Archytas are lost.[22] The writings of Aristotle's student Aristoxenus (a native of Tarentum) are understood as central to the later biographical works, such as those of Diogenes Laertius.

What is known with reasonable certainty is that Archytas was an accomplished mathematician who solved the geometrical problem of doubling a cube (the famous "Delian problem") and is thought to have been the first to use mathematics in the study of mechanics. In a surviving fragment from one of his mathematical works, he seems to affirm Philolaus's conviction that number and calculation are key to understanding nature:

> Those concerned with the sciences seem to me to make distinctions as well, and it is not at all surprising that they have correct

---

21   Ibid., 4.

22   This causes me no small amount of grief. The texts written by Aristoxenus (a native of Tarentum) are understood as central to the later biographical works, such as those of Diogenes Laertius, from which accounts and fragments of Archytas are drawn. See Huffman, *Archytas of Tarentum*, 4. Huffman later notes (*Archytas of Tarentum*, 19) that the work by Aristoxenus has a panegyric quality to it but argues that there are good reasons for taking it as a largely reliable source.

understanding about individual things as they are. For, having made good distinctions concerning the nature of wholes they were likely also to see well how things are in their parts. Indeed, concerning the speed of the stars and their risings and settings as well as concerning geometry and numbers and not least concerning music, they handed down to us a clear set of distinctions. For these sciences seem to be akin.[23]

Notice that Archytas names all four of the mathematical arts that would later come to be known collectively as the *quadrivium*: astronomy, geometry, arithmetic, and music (harmonics)—the mathematical curriculum Plato outlines in *The Republic*, where he refers to these disciplines as "kindred sciences."[24] In a passage written by Eudemus of Rhodes (a student of Aristotle) that was preserved by the neo-Platonist Proclus, Archytas is cited as one of the three prominent geometers of Plato's generation.[25] Moreover, Archytas was the foremost harmonic theorist of his time; even the great mathematician Ptolemy (second century AD), in his *Harmonics*, alluded to Archytas's highly sophisticated work with musical scales and their associated mathematical ratios. Huffman concludes, "Archytas of Tarentum fits the popular conception of a Pythagorean better than anyone in the Pythagorean tradition, including Pythagoras himself."[26]

---

23 Guthrie, *A History of Greek Philosophy*, 336. This fragment is regarded as genuine by Huffman, et al.

24 Plato, *The Republic*, VI.510.

25 Huffman, *Archytas of Tarentum*, 46.

26 Ibid., 44.

Pythagoreanism According to Aristotle

The testimony of Aristotle offers important insights about the Pythagorean tradition as he knew it and serves as valuable support for the authenticity of certain fragments and testimonia of Philolaus and Archytas. As a disciple of Plato, Aristotle was certainly able to distinguish Pythagorean thought of the fourth century BC from the Platonized versions that may have emerged in the early Academy. As Guthrie explains,

> we must do all we can to distinguish between Pythagoreanism up to Plato's time and the philosophy of Plato himself, which certainly owed much to it, and which tended to be read back into Pythagoreanism by its contemporaries and successors. For this purpose no guide can be as good as Aristotle, since the man who was a member of the Academy for twenty years of Plato's lifetime certainly knew the difference between the two and refers to it more than once.[27]

Put simply, Aristotle's writings on Pythagoreanism help scholars roughly identify the beginning of the hybridized philosophy of the Pythagorean-Platonic tradition, a delineation that is important to the project of intellectual history. Fortunately, Aristotle mentions Pythagorean ideas multiple times in different surviving works and even contrasts the philosophy with Platonism in one passage, which provides particularly useful insight.

In Book I of the *Metaphysics*, Aristotle writes that the Pythagoreans (the first mathematicians) regarded numbers as fundamental to physical reality—"the substance of all things"—and mathematics as the "principles of all things."[28] In Book XIII, he discusses

---

27   Guthrie, *A History of Greek Philosophy*, 215.

28   Aristotle, *Metaphysics*, I.987a, I.985b. It has been argued that Aristotle misread Philolaus on this point. See Endnote 4.

the Pythagorean view that sensible substances are numbers made manifest in space: "For they construct the whole universe out of numbers—only not numbers consisting of abstract units; they suppose the units to have spatial magnitude."[29] Geometry (number in space), harmonics (number in time), and astronomy (number in space and time) were understood as interconnected aspects of reality that produced a cosmic harmony—the music of the spheres:

> ...they saw that the modifications and the ratios of the musical scales were expressible in numbers;—since, then, all other things seemed in their whole nature to be modelled on numbers, and numbers seemed to be the first things in the whole of nature, they supposed the elements of numbers to be the elements of all things, and the whole heaven to be a musical scale and a number. And all the properties of numbers and scales which they could show to agree with the attributes and parts and the whole arrangement of the heavens, they collected and fitted into their scheme; and if there was a gap anywhere, they readily made additions so as to make their whole theory coherent.[30]

One example Aristotle gives of an *ad hoc* addition the Pythagoreans used to justify an *a priori* belief about the cosmos is the postulation of an unobservable "counter-earth" that brought the total number of heavenly bodies (which also included the earth, moon, sun, five visible planets and sphere of fixed stars) to ten—the sacred number of perfection represented by the triangular tetractys.

---

29   Aristotle, *Metaphysics*, XIII.1080b.

30   Ibid., I.985b–986a.

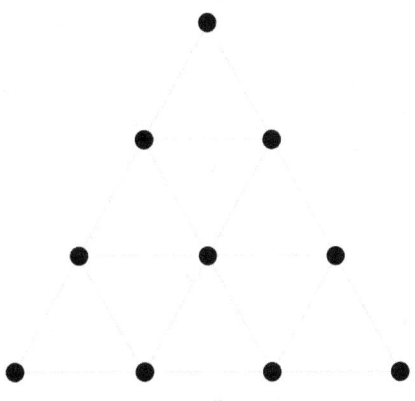

Figure 1.2 *The Tetractys*

He discusses this "philosophy first" approach to cosmology in *On the Heavens*:

> At the centre, they say, is fire, and the earth is one of the stars, creating night and day by its circular motion about the centre. They further construct another earth in opposition to ours to which they give the name counter-earth. In all this they are not seeking for theories and causes to account for observed facts, but rather forcing their observations and trying to accommodate them to certain theories and opinions of their own…their view is that the most precious place befits the most precious thing: but fire, they say, is more precious than earth…Reasoning on this basis they take the view that it is not earth that lies at the centre of the sphere, but rather fire.[31]

This fire-centric, ten-body model was rejected by Aristotle, who postulated that the earth sits at the center of a system of nested, rotating spheres in which the planets, moon, and sun are embedded. Nevertheless, these passages are of significant value in that

---

31 Aristotle, *On the Heavens*, II.293a–293b.

they reveal the character of late fifth- and early fourth-century Pythagoreanism.

Another way Aristotle benefits this investigation is by explicitly contrasting Pythagorean thought with Platonism. Caution is in order, however, since Aristotle seems to impose some measure of philosophical bias in his treatment of the subject.[32] In Book I of the *Metaphysics*, he discusses Plato's understanding of the relationship between number and the sensible world and how this view compares to that of previous thinkers, including the Pythagoreans. Aristotle writes that Plato regarded the world as a dichotomy of sensible things and abstract Forms (ideas) in which sensible things "participate": "Only the name 'participation' was new," explains Aristotle, "for the Pythagoreans say that things exist by 'imitation' of numbers, and Plato says they exist by participation, changing the name. But what the participation or the imitation of the Forms could be they left an open question."[33] In other words, Aristotle equates the Pythagorean and Platonic understanding of the relationship between immaterial number and material reality.[34] He explains that Plato saw "objects of mathematics" not as Forms themselves, but as intermediary entities "differing from sensible things in being eternal and unchangeable, from Forms in that there are many alike, while the Form itself is in each case unique."[35] It has been suggested, however, that Plato eventually came to see the Forms as mathematical entities of some kind.[36] Whatever the case may be, Aristotle understood Plato's doctrine of Forms as an extension of Pythagorean thought but believed that Plato differed

---

32   See Guthrie, *A History of Greek Philosophy*, 215, for elaboration on this apparent bias.

33   Aristotle, *Metaphysics*, I.987b.

34   Aristotle may have failed to fully appreciate Platonic nuance on this matter. See Endnote 5.

35   Ibid.

36   John Dillon, *The Middle Platonists: 80 BC to AD 220* (Ithaca: Cornell University Press, 1996), 4.

from the Pythagoreans in his view that numbers are ontologically distinct from sensible things, not the essence of things themselves.[37] In other words, Plato would not agree that "all is number"; rather, he saw numbers as an intermediary between the Forms and their imperfect material copies.

Plato's *Timaeus* as a Key Transmitter of Pythagorean Thought

The only Platonic writing to which European scholars had direct access prior to the late Middle Ages was the *Timaeus*, which was included in the liberal arts curriculum and was crucial in the transmission of Pythagorean-Platonic tradition to the Christian West. The dialogue is a poetic work of philosophical myth featuring a Pythagorean character, Timaeus, who tells a "likely story" (i.e., a plausible account, the best we can hope to achieve) of the formation of the cosmos—what is now referred to as a cosmogony. Significantly, Plato departs from Pythagorean thought by positing a good Craftsman, an eternal Father who predates matter itself and who shaped the world according to a preexisting model.[38] The mathematical structure of the cosmos is a central theme of Timaeus's creation story; he explains that the four elements thought to be fundamental to the sensible realm acquired their essential character by having a rational order imposed upon them:

---

37   Aristotle, *Metaphysics*, I.987b.

38   Louis Markos, *From Plato to Christ: How Platonic Thought Shaped the Christian Faith* (Downers Grove: IVP Academic, 2021), 102–103. Markos writes, "Timaeus is the only ancient book to posit a Creator who *predates* matter…though Plato does not specifically say that his Demiurge created the world out of nothing (*ex nihilo*), the *Timaeus* is the only book to come within a thousand miles of this central biblical claim…Plato is the only non-Jewish ancient writer to suggest, if not clearly state, that god is the absolute origin of *all* things, whether physical or spiritual, temporal or eternal."

> Wherefore...the various elements had different places before they were arranged so as to form the universe. At first, they were all without reason and measure. But when the world began to get into order, fire and water and earth and air had only certain faint traces of themselves, and were altogether such as everything might be expected to be in the absence of God; this, I say, was their nature at that time, and God fashioned them by form and number.[39]

He then describes the sequential geometric construction of the tetrahedron (pyramid), octahedron, icosahedron, and cube, and assigns each of these to one of the four elements:

> To the earth, then, let us assign the cubical form; for earth is the most immoveable of the four...Let it be agreed, then, both according to strict reason and according to probability, that the pyramid is the solid which is the original element and seed of fire; and let us assign the element which was next in the order of generation [the octahedron] to the air, and the third [the icosahedron] to water.[40]

In this way, the four basic elements were "fashioned by form and number" and then the world was "harmonized by proportion" so that it was the best world possible, given any inherent limitations of the material substrate.[41] The fifth of the so-called Platonic solids—the twelve-faced dodecahedron—is alluded to and assigned to the cosmos as a whole: "There was yet a fifth combination which God used in the delineation of the universe."[42]

---

39   Plato, *Timaeus*, 53b.

40   Ibid.

41   Ibid., 53b, 32c.

42   Ibid., 55c. In Plato's *Phaedo* (110b), Socrates mentions a ball made from twelve pieces of leather stitched together, which seems to refer to the dodecahedron. In that context, he says that this is the shape of the earth (not to be confused with elemental earth, which is cubical). For more on Plato's geometrical cosmos, see Endnote 6.

THE FIVE PLATONIC SOLIDS

TETRAHEDRON   CUBE   OCTAHEDRON   DODECAHEDRON   ICOSAHEDRON

Figure 1.3 *The Five Platonic Solids*

In order to set everything into motion, the world was infused with soul, becoming an animated creature permeated by rationality and design. The Craftsman even created time itself; the system of heavenly bodies is a "moving image of eternity" that operates according to number and tracks time's passage by signifying days, nights, months, and years.[43] Timaeus explains that the material world and time are inextricably connected: "Time, then, and the heavens came into being at the same instant in order that, having been created together, if ever there was to be a dissolution of them, they might be dissolved together."[44]

As previously discussed, the Pythagoreans related the mathematical orderliness imposed by the limiters to the possibility of human knowledge of the natural world. Similarly, Timaeus draws a connection between the inherent rationality of the world and the capacities of the human mind, a relationship that renders the cosmos intelligible. He says, "God lighted a fire, which we now call the sun…that it might give light to the whole of heaven, and that the animals, as many as nature intended, might participate in number,

---

43   Plato, *Timaeus*, 37d.

44   Ibid., 38b. The similarity of this idea and the contemporary conception of space-time is remarkable. For more on Plato's view of the temporality of matter, see *Timaeus*, 28b.

learning arithmetic from the revolution of the same and the like."[45] Plato scholar T. K. Johansen points out that "we see a kind of anthropocentricity…in the view that the sun illuminates the heavens so that by observing the planets 'those animals to which it was appropriate' can learn the mathematical regularities that govern their motions."[46] Timaeus explains that the human soul was created with the sensory and intellectual faculties necessary for the appreciation of the rational order and harmony made manifest in the cosmos:

> …the sight of day and night, and the months and the revolutions of the years, have created number, and have given us a conception of time, and the power of enquiring about the nature of the universe; and from this source we have derived philosophy, than which no greater good ever was or will be given by the gods to mortal man…God invented and gave us sight to the end that we might behold the courses of intelligence in the heaven, and apply them to the courses of our own intelligence which are akin to them… and that we, learning them and partaking of the natural truth of reason, might imitate the absolutely unerring courses of God…[47]

In the cyclical regularities of the heavens and their marking of time's progression, Timaeus perceives an orderliness that is mirrored in his own soul. As classicist Charles Cochrane puts it, the mind of man is "identified with the cosmic principle and conceived as a 'scintilla' of the divine essence; that is why…it is held to be capable of apprehending the archetypal forms."[48] Through its resonance with the "courses of intelligence" in the heavens—which we

---

45 Plato, *Timaeus*, 39b–39c.

46 T. K. Johansen, *Plato's Natural Philosophy: A Study of the Timaeus-Critias* (Cambridge: Cambridge University Press, 2008), 3.

47 Plato, *Timaeus*, 47a–47c.

48 Charles Cochrane, *Christianity and Classical Culture* (Indianapolis: Oxford University Press, 1940), 88.

are able to behold using our faculty of sight—the soul is inspired to contemplate the transcendent.

In its structure, Plato's cosmology is quite different from the Pythagorean in that there is no central fire or counter-earth, and the sun, moon, and five visible planets revolve around our spherical observational platform. However, Plato preserves the idea of the music of the spheres, describing the revolutions of the heavenly bodies as proportional to musical intervals, using a geometrically derived scale rather than the arithmetical Pythagorean scale.[49] He emphasizes the need for both intelligent causation and necessity in the formation and operation of the cosmos: "Both kinds of causes should be acknowledged by us, but a distinction should be made between those which are endowed with mind and are the workers of things fair and good, and those which are deprived of intelligence and always produce chance effects without order or design."[50] By "workers of things fair and good" he seems to mean the Craftsman's use of the rational Forms as a pattern for the fabrication of the visible world, and by causes "deprived of intelligence" he seems to mean blind, mechanical causes that operate by physical necessity.

Thus, with the *Timaeus*, the Pythagorean-Platonic tradition was born.

---

[49] Andrew Gregory, "The Pythagoreans: Number and Numerology," in *Mathematicians & Their Gods* edited by Snezana Lawrence and Mark McCartney (Oxford: Oxford University Press, 2015), 42.

[50] Plato, *Timaeus*, 46e.

CHAPTER 2

Developments in Middle Platonism

Middle Platonism is the name given to the period of Western intellectual history that began around 80 BC and ended around AD 220. According to historian of philosophy John Dillon, "It was during this period...that much of what passed for Platonism in later ages—until a determined effort was made at the beginning of the nineteenth century in Germany to return to the actual writings of Plato, unvarnished by interpretation—was laid down."[1] There was a renewed interest in Pythagoreanism, what Dillon refers to as "one of the more noticeable intellectual developments of the first century BC."[2] This so-called neo-Pythagoreanism was largely based upon an emergent body of pseudepigraphal writings that blended Platonic and Aristotelian teachings and painted them as doctrines that sprang from Pythagoreanism.[3] Neo-Pythagoreanism was syncretized with the redeveloped Platonism of the time, and the resulting amalgam is what came to be known as Middle Platonism. Two figures of this

---

1   John Dillon, *The Middle Platonists*, 1.

2   Ibid., 117.

3   Dillon describes these highly questionable works thus: "The treatises are bald and didactic, stating their doctrine without attempt at proof, and aimed at an audience which, it would seem, was prepared to substitute faith for reason" (Dillon, *The Middle Platonists*, 119). Note that some writers use a spelling variation of neo-Pythagoreanism and neo-Pythagorean, but for the sake of uniformity and clarity, these are the forms used throughout this book.

period are of particular interest: Eudorus of Alexandria for his key metaphysical contribution, and Philo of Alexandria (whose immediate philosophic source was Eudorus) for his integration of Platonism with monotheism and the enormous influence of his writings upon the early Christian Platonists.

The enigmatic Eudorus flourished sometime during the middle of the first century BC and likely acquired his Platonism through his Alexandrian education. Much of what is known about his philosophy is gleaned from fragments and allusions and by careful deduction from the spirit of his references to other philosophers. Only one of his works is known by title, *A Concise Survey of Philosophy*, a summary of which is preserved in Arius Didymus.[4] It is possible, though not certain, that Eudorus penned a *Life of Pythagoras* that was fairly influential.[5] Either way, Eudorus clearly represents a new influence in Middle Platonism—a desire to recover traces of genuine Pythagoreanism, particularly by way of the writings of Philolaus and Archytas, but also using accounts provided by fourth-century Platonists and Peripatetics, including Aristotle and Aristoxenus.[6] It is known that Eudorus commented on Plato's *Timaeus*, which was instrumental in sparking a revival of interest in Pythagoreanism in philosophical circles. Plutarch, in *On the Generation of the Soul in the Timaeus*, cites this commentary three different times: once in a discussion of the generation of the world and soul, and twice in discussions of harmonic ratios.[7] Available evidence suggests that Eudorus's cosmology was likely of the Stoic variety, which means he understood the heavens to be a realm of

---

4 Dillon, *The Middle Platonists*, 116.

5 Walter Burkert, *Lore and Science in Ancient Pythagoreanism*, 53.

6 Dillon, *The Middle Platonists*, 119.

7 See Plutarch, *Moralia, Volume XIII: Part 1: Platonic Essays*, translated by Harold Cherniss (Cambridge, MA: Harvard University Press, 1976), 171, 295, 301.

pure fire containing divine bodies constituted of pure fire, with harmonic intervals between the spheres.[8]

One of the major developments of Middle Platonism that can reasonably be credited to Eudorus is the identification of the One, a "supreme god" who is the transcendent causal principle of matter and all created things.[9] Subordinate to the One are the Monad and Unlimited Dyad, opposing principles existing in duality (what the earlier Pythagoreans referred to as the unlimited and limiters). In a text preserved by Simplicius in a commentary on Aristotle's *Physics*, Eudorus writes that "the One is first principle of all things, since matter and all Beings [the Forms] have come into being from it. And this is the supreme god."[10] It is notable that the god postulated by Eudorus is the grounding of both matter and the Platonic Forms (which he seems to have understood as numbers).[11]

Eudorus's conception of the soul is not made explicit in the surviving evidence, but it can be reasonably inferred. It is clear that he had a favorable view of the philosophy of Xenocrates, who was head of Plato's Academy beginning in 339/338 BC. Xenocrates's discussion of the soul occurs in *On Nature*, a work that only survives in fragments. His pupil, Crantor, emphasized the importance of the soul's ability to relate to both the immaterial, intelligible world and the material world, "a circumstance," says Dillon, "which required it to be compounded of elements of both."[12] This is a fine example of the idea that an ontological *and* epistemological resonance exists between the soul of man and the order of the visible cosmos, which is modeled after the Platonic Forms. What we have here is an interconnection between a transcendent creator,

---

8   Dillon, *The Middle Platonists*, 130.

9   Ibid., 126–127.

10 Simplicius, *In Phys.* 181, as quoted in Kahn, *Pythagoras and the Pythagoreans*, 97.

11 Dillon, *The Middle Platonists*, 129.

12 Ibid., 131.

matter, number, and the human soul (mind). Turning now to Philo, a Middle Platonist and intellectual successor of Eudorus, we shall see how this line of thought was expanded and harmonized with monotheism.

Philo Judaeus (c. 20 BC–AD 50) has been described as "one of the most remarkable literary phenomena of the Hellenistic world."[13] He came from an affluent family of the Alexandrian Jewish Diaspora, and his writings constitute the largest corpus from such a Jew. He was both a philosopher of the Pythagorean-Platonic tradition and a theologian concerned with precise interpretation of the Pentateuch; he said, "I venture not only to study the sacred commands of Moses, but also with an ardent love of knowledge to investigate each separate one of them, and to endeavour to reveal and to explain to those who wish to understand them, things concerning them which are not known to the multitude."[14] This fusion of Hellenistic philosophy and Jewish monotheism in the writings of a Middle Platonist is what makes the Philonic corpus an incomparable treasure trove. As one biographer explains:

> Philo stood at the crossroads of a number of ancient philosophical movements. He consistently drew on Stoic, Platonic, and Pythagorean traditions in order to interpret the Jewish Scriptures. Thus he provides us with an "archaeological dig" from which we can probe the development of ancient philosophy. Not only do his writings preserve valuable philosophical traditions prior to him, but they also give us a peek at the seeds from which neo-Platonism sprang in the third century CE.[15]

Fortunately, Philonic scholarship has an abundance of source material; much of Philo's work has survived intact, thanks to Christian

---

13   Dillon, *The Middle Platonists*, 139. Also known as Philo of Alexandria.

14   Philo, *The Special Laws*, III.1.

15   Kenneth Schenck, *A Brief Guide to Philo* (Louisville: John Knox Press, 2005), 3.

thinkers who considered it to be an invaluable reference point for the New Testament (which also came out of Hellenistic Judaism).[16]

It is well established that Philo's education followed the Greek *enkyklios paideia* and that he understood philosophy as the pursuit of wisdom in service to theology. In *On Mating with the Preliminary Studies*, he argues that wisdom has its own servant: the "encyclical knowledge" that includes logic, music, rhetoric, grammar, geometry, astronomy, "and all the other sorts of contemplation which proceed in accordance with reason; of which Hagar, the handmaid of Sarah, is an emblem."[17] It is through these studies, he says, that the soul must first travel on its quest for its true "food":

> Do you not see that our bodies do not use solid and costly food before they have first, in their age of infancy, used such as had no variety, and consisted merely of milk? And, in the same way, think also that infantine food is prepared for the soul, namely the encyclical sciences, and the contemplations which are directed to each of them; but that the more perfect and becoming food, namely the virtues, is prepared for those who are really full-grown men.[18]

All the sciences that make up what he refers to as "preliminary studies" are meant to cultivate a soul that, upon maturity, feeds upon virtue.

Philo was steeped in Plato, whom he called "the sweetest of all writers," and seems to have had a particular attraction to the *Timaeus*; his corpus includes twenty instances of either paraphrase, quotation, or direct reference to that dialogue.[19] Moreover, his adoption of imagery and use of quoted passages suggest a thorough knowledge of the text. The flavor of Platonism is immedi-

---

16 Ibid., 2–3.

17 Philo, *On Mating with the Preliminary Studies*, III.

18 Ibid., IV.

19 David Runia, *Philo of Alexandria and the Timaeus of Plato* (Leiden, The Netherlands: E. J. Brill, 1986), 367, 371.

ately detectable in Philo's discussion of the mathematical sciences. He says that

> music will teach what is harmonious...and, rejecting all that is out of tune and all that is inconsistent with melody, will guide what was previously discordant to concord. And geometry, sowing the seeds of equality and just proportion in the soul...will, by means of the beauty of continued contemplation, implant in you an admiration of justice."[20]

He views the cosmos as "the most excellent of all created things" and astronomy as the "queen of all the sciences."[21] Through the study of astronomy, one "is allowed to see that which is second best [to seeing God himself], namely, the heaven which is perceptible by the external senses, and the harmonious arrangement of the stars therein, and their truly musical and well-regulated motion."[22] In accord with the Pythagorean-Platonic tradition, Philo sees number as that which makes the creation of the physical world as well as its comprehensibility possible.[23]

In his cosmogony, which he outlines in *On the Creation*, Philo takes the biblical creation account to be true, yet makes liberal use of Pythagorean-Platonic themes in his exegesis. More than a quarter of the text is concerned with the significance of the different numbers mentioned in the Genesis narrative, a fact that is indicative of his belief that created things are ordered, and that "number is inherent in order."[24] An interesting example is his explanation for the creation of the heavenly bodies on the fourth day of the creation week. He takes the number four in this context to repre-

---

20 Philo, *On Mating with the Preliminary Studies*, IV.

21 Ibid., X.

22 Ibid.

23 Charles Kahn, *Pythagoras and the Pythagoreans*, 101.

24 David Runia, *On the Creation of the Cosmos According to Moses: Introduction, Translation and Commentary* (Atlanta: Society of Biblical Literature, 2001), 26.

sent the first four natural numbers, which comprise the Pythagorean tetractys (1+2+3+4=10). He writes:

> And next the heaven was embellished in the perfect number four, and if any one were to pronounce this number the origin and source of the all-perfect decade [ten] he would not err. For what the decade is in actuality, that the number four, as it seems, is in potentiality, at all events if the numerals from the unit to four are placed together in order, they will make ten, which is the limit of the number of immensity, around which the numbers wheel and turn as around a goal.[25]

He goes on to relate the number four to "the principles of the harmonious concords in music" (the ratios associated with harmonic theory) and with geometry. He writes, "There is also another power of the number four which is a most wonderful one to speak of and to contemplate. For it was this number that first displayed the nature of the solid cube, the numbers before four being assigned only to incorporeal things."[26] He explains that the number one represents the geometrical point, which has no length or breadth; the number two represents the line, which has length but no breadth; the number three represents plane (two-dimensional) figures, which have length and breadth but no depth; and four represents the figures that have spatial extension in three dimensions. Like the Platonic scheme, Philo's cosmology is geocentric, with the earth "in the most centre spot of the universe," but he regards the sun's place as the most important.[27] In a work entitled *Who is the Heir of Divine Things*, he explains that the sun is (in a sense) at the "center" of the planets, with Saturn, Jupiter, and Mars revolving "above" and Mercury, Venus, and the moon revolving "below."[28]

---

25 Philo, *On the Creation*, XV.
26 Ibid., XVI.
27 Philo, *On the Life of Moses*, I.XXXVIII.
28 Philo, *Who is the Heir of Divine Things*, XLV.

Philo agrees with his intellectual predecessors that the cosmos was created according to a rational plan, but insists that this plan was grounded in the mind of God:

> God, as apprehending beforehand, as a God must do, that there could not exist a good imitation without a good model, and that of the things perceptible to the external senses nothing could be faultless which was not fashioned with reference to some archetypal idea conceived by the intellect, when he had determined to create this visible world, previously formed that one which is perceptible only by the intellect, in order that so using an incorporeal model formed as far as possible on the image of God, he might then make this corporeal world, a younger likeness of the elder creation, which should embrace as many different genera perceptible to the external senses, as the other world contains of those which are visible only to the intellect.[29]

Here, Philo underscores the Creator's sovereignty over creation by insisting that the immaterial blueprint—the archetypal idea—for the construction of the world was a preliminary product of the divine intellect. As David Runia puts it, "The promotion…is striking. Philo is underlining not only the complexity of the cosmic megalopolis, but also the superior status of the architect who does not merely copy an already existing model but designs the plan himself."[30] Further on in the same work, Philo reiterates the idea that the rational blueprint for the cosmos was instituted prior to its corporeal manifestation:

> God, who having determined to found a mighty state, first of all conceived its form in his mind, according to which form he made a world perceptible only by the intellect, and then completed one visible to the external senses, using the first as a model. As therefore the city, when previously shadowed out in the mind of the

---

29 Philo, *On the Creation*, IV.

30 David Runia, *Philo of Alexandria and the Timaeus of Plato*, 421.

man of architectural skill had no external place, but was stamped solely in the mind of the workman, so in the same manner neither can the world which existed in ideas have had any other local position except the divine reason which made them...[31]

"Philo believed that he saw in Genesis the description of a double creation," explains Dillon, "first that of the intelligible work (*noêtos kosmos*), then that of the sensible world (*aisthêtos kosmos*)."[32] Philo regards this noetic cosmos as the "archetypal seal" or "Reason of God" (*Logos*) that is stamped upon both mankind—God's image-bearers—and the world itself.[33] The *Logos* informs the tangible world, accounting for its orderly structure as well as man's rational faculties—it "imposes intelligibility and order on brute physicality."[34] Since the Reason of God is reflected (to a limited degree) in mankind—the intellect being the "divine part" of a human being—we are predisposed to recognize signs of the *Logos* in both our own minds and the rest of creation. In other words, it is our rational kinship with both the Creator and the cosmos that allows us to recognize His archetypal seal.

It is important to understand that Philo was probably not the first philosopher to explicitly situate the archetype for creation in the mind of God. Antiochus of Ascalon (c. 130–69 BC), another Middle Platonist, seems to have viewed the Forms as thoughts of God, and it is considered possible (though not provable) that this conception goes back as far as some philosophers of the original Academy.[35] Also recall that Eudorus, who lived during the first century BC, grounded the Platonic Forms in the One (the supreme God). What is abundantly clear is that Philo attributed the observ-

---

31 Philo, *On the Creation*, IV–V.
32 Dillon, *The Middle Platonists*, 158.
33 Philo, *On the Creation*, VI.
34 Runia, *On the Creation of the Cosmos According to Moses*, 16.
35 Dillon, *The Middle Platonists*, 95.

able order of the universe to rational causation and recognized the excellent compatibility between Greek thought and Jewish cosmology. Perhaps he saw the harmony between the Greek concept of *Logos* and the Hebrew notion of the creative "word" of God, as illustrated in Psalm 33:6: "By the word of the Lord the heavens were made, and by the breath of his mouth all their host." Philo grafted his Hellenistic philosophy into his theology of creation in a way that served to support and illuminate Jewish Scripture. The result was a richer picture of reality than philosophy alone provided: one of a personal God Whose mind is reflected in His creative works and is thereby perceived by human beings, the only creatures bearing His image.

Another work of Middle Platonism that should be mentioned is *The Wisdom of Solomon*, a piece of Jewish wisdom literature that synthesizes Greek and Jewish thought in a manner quite reminiscent of Philo. David Winston, a specialist in rabbinic and Hellenistic literature, dates this writing to sometime after the advent of the Roman period (30 BC), which means that the author could have been an Alexandrian contemporary of Philo.[36] Among other themes of Middle Platonism, *Wisdom* contains a brief yet striking passage on God's work of creation and its mathematical orderliness. In the eleventh chapter, the author addresses God saying, "you ordered all things by measure and number and weight."[37] It is significant that a book about the wisdom of God uses language associated with number to describe the mindful arrangement of nature. As will be seen, this idea would go on to be adopted by prominent Christian thinkers, particularly those well versed in Greek philosophy.

In a manner resonant with Philo's work, *Wisdom* affirms the legitimacy of natural revelation as such by celebrating the order

---

36 David Winston, *The Wisdom of Solomon: A New Translation with Introduction and Commentary* (New York: Doubleday, 1979), 3.

37 Ibid., 230.

and beauty of nature and man's ability to discern it.[38] At one point, the writer asks how some, who "are not to be excused," can perceive the order and harmony of the cosmos yet fail to recognize the Artificer: "For from the greatness and beauty of created things, is their author correspondingly perceived."[39] This passage bears a striking resemblance to the words of the Apostle Paul to the Roman church: "For his invisible attributes, namely, his eternal power and divine nature, have been clearly perceived, ever since the creation of the world, in the things that have been made. So they are without excuse."[40] It is not much of a stretch to suppose that Paul, a highly educated Jew and Roman citizen, was familiar with (and perhaps influenced by) *The Wisdom of Solomon*.

Nicomachus of Gerasa

We will conclude this selective survey of the Middle Platonists with Nicomachus of Gerasa[41] (fl. c. AD 100), a mathematician and philosopher who made considerable contributions to the ongoing Pythagorean-Platonic tradition and the liberal arts curriculum. His *Introduction to Arithmetic* enjoyed great success as a textbook; it was widely used in later antiquity and Boethius's Latin paraphrased version was used throughout the Middle Ages.[42] Nicomachus also

---

38 For a full accounting of the similarities between Philo and *The Wisdom of Solomon*, see the cross-references in Winston.

39 Winston, *The Wisdom of Solomon*, 247.

40 Romans 1:20.

41 This Gerasa was most likely in Palestine and was primarily Greek. See Mortimer Adler, "Biographical Note: Nicomachus," Great Books of the Western World 10 (Chicago: Encyclopaedia Britannica, Inc., 1990), 595.

42 Nicomachus of Gerasa, *Introduction to Arithmetic*, Great Books of the Western World 10 (Chicago: Encyclopaedia Britannica, Inc., 1990), 596.

wrote an *Introduction to Harmonics*[43] and a *Theology of Arithmetic*[44] that have survived, and (among other things) an *Introduction to Geometry* and a *Life of Pythagoras* that have been lost.[45] As a neo-Pythagorean, he perpetuated some of the popular lore of Pythagorica, such as the claim that Pythagoras systematized science and defined philosophy as the love of wisdom.[46] Thus, he was a key transmitter of the shade of Pythagoreanism that carried into the Middle Ages and beyond.[47]

Nicomachus may be technically classified as a neo-Pythagorean, but as Dillon points out, "his philosophy fits comfortably within the spectrum of contemporary Platonism."[48] For example, very early on in his *Introduction to Arithmetic*, Nicomachus makes the distinction between things of the intelligible realm and things of the sensible realm. In describing the Forms he writes: "Those things…are immaterial, eternal, without end, and it is their nature to persist ever the same and unchanging, abiding by their own essential being, and each one of them is called real in the proper sense."[49] By contrast, he says, the things of the sensible realm are only real insofar as they participate in the intelligible things, "for they do not abide for even the shortest moment in the same condi-

---

43 Charles Kahn notes that the *Harmonics* was the only treatise on harmonics from the period between Euclid and Ptolemy to survive in its entirety. See Kahn, *Pythagoras and the Pythagoreans*, 111.

44 Also known as *Arithmetical Theology*, this work is partially preserved in a text known as *Theology of Arithmetic* attributed to Iamblichus. It was important to Proclus, who will be discussed in Chapter 3.

45 Nicomachus, *Introduction to Arithmetic*, 595.

46 Ibid., 599.

47 Proclus, an important neo-Platonist who had a significant influence on Kepler, claimed (in true Pythagorean fashion) to be the reincarnation of Nicomachus. See A. H. Criddle, "The Chronology of Nicomachus of Gerasa" in *The Classical Quarterly* 48, no. 1 (1998), 324 n. 2.

48 Dillon, *The Middle Platonists*, 353.

49 Nicomachus, *Introduction to Arithmetic*, 599.

tion, but are always passing over in all sorts of changes."⁵⁰ He then quotes a relevant passage from the *Timaeus*: "What is that which always is, and has no birth, and what is that which is always becoming but never is? The one is apprehended by the mental processes, with reasoning, and is ever the same; the other can be guessed at by opinion in company with unreasoning sense, a thing which becomes and passes away, but never really is."⁵¹ Like earlier thinkers of the Pythagorean-Platonic tradition, he sees the necessity of number for illuminating intelligible and sensible reality. Citing Pythagorean philosopher Androcydes, he argues that without the aid of the mathematical sciences,

> it is not possible to deal accurately with the forms of being nor to discover the truth in things, knowledge of which is wisdom, and evidently not even to philosophize properly, for "just as painting contributes to the menial arts toward correctness of theory, so in truth lines, numbers, harmonic intervals, and the revolutions of circles bear aid to the learning of the doctrines of wisdom…⁵²

Notably, Nicomachus then quotes Archytas in naming the four mathematical disciplines in his discussion of the value of these "sister sciences":

> Archytas of Tarentum, at the beginning of his treatise *On Harmony*, says… 'It seems to me that they [the Pythagoreans] do well to study mathematics, and it is not at all strange that they have correct knowledge about each thing, what it is. For if they knew rightly the nature of the whole, they were also likely to see well what is the nature of the parts. About geometry, indeed, and arithmetic and astronomy, they have handed down to us a clear understanding, and not least also about music. For these seem to

---

50 Ibid.

51 Ibid. This quotation of the *Timaeus* is taken from section 27 of that dialogue.

52 Nicomachus, *Introduction to Arithmetic*, 600.

be sister sciences; for they deal with sister subjects, the first two forms of being."[53]

Although he obviously values all four of the mathematical sciences, he places arithmetic in the first place of importance in the progression of learning; he calls it the "more honorable, and more venerable, and, as it were, mother and nurse of the rest."[54]

In a particularly striking passage, Nicomachus describes his mathematical cosmology:

> All that has by nature with systematic method been arranged in the universe seems both in part and as a whole to have been determined and ordered in accordance with number, by the forethought and the mind of him that created all things; for the pattern was fixed, like a preliminary sketch, by the domination of number preexistent in the mind of the world-creating God, number conceptual only and immaterial in every way, but at the same time the true and the eternal essence, so that with reference to it, as to an artistic plan, should be created all these things, time, motion, the heavens, the stars, all sorts of revolutions.[55]

The kinship with Philonic thought here is evident; Nicomachus sees the cosmos as mathematically ordered according to an archetype that first existed in the intellect of the creator. Additionally, in a passage of the *Harmonics*, he connects the ratios of music theory to the motions of the heavenly bodies, thereby preserving the traditional Pythagorean view. Moreover, he reinterprets the Forms in Pythagorean style: as numbers themselves.[56]

One thing that is uncertain in Nicomachus's philosophy is what exactly Nicomachus means by "God," whom he equates with

---

53   Nicomachus, *Introduction to Arithmetic*, 600.

54   Ibid., 601.

55   Ibid., 602.

56   Kahn, *Pythagoras and the Pythagoreans*, 114.

the Monad in his *Theology of Arithmetic*.⁵⁷ Dillon believes that there is no detectable distinction made between a supreme God and an impersonal, demiurgic one. It could be that for Nicomachus, the Demiurge *is* the highest God; the identification with the Monad makes this interpretation plausible.⁵⁸ Also related to this question is Nicomachus's understanding of the *Logos*, which he describes as the "offspring" of the Monad and the Dyad. In any case, the Nicomachean system reflects the Platonic triad of God, matter, and Form (*Logos*), which is a central theme of this study.⁵⁹

---

57 Dillon, *The Middle Platonists*, 355.
58 Ibid., 355, 357.
59 Ibid., 358.

CHAPTER 3

The Pythagorean-Platonic Tradition from Ptolemy to Proclus

The Pythagorean-Platonic tradition persisted into late Antiquity and then the Middle Ages with the work of philosophers and theologians known to have been significantly influential upon Kepler's work—or at least prominent in the intellectual milieu that shaped him. These included Ptolemy, Augustine, Proclus, Nicholas of Cusa, and Nicolaus Copernicus, among others, who were vital to the development of his natural philosophy.[1] As this selective survey of intellectual history continues, it should be noted that it is not known, in every case, which *specific* works Kepler read for himself; the best that can be done in such instances is to canvas the work of the referenced writer for passages that are thematically consonant. We shall begin with the later years of Middle Platonism and proceed through the work of Nicolaus Copernicus, who was, at least in some respects, Kepler's most important predecessor.

---

1 Max Caspar writes, "Consciously or unconsciously, Kepler's thoughts were connected with everything which he had heard and read of Pythagoras and Plato, of Augustine and Nicholas of Cusa and many other great men of the past and with that which Christian teaching about God and the world and the position of men regarding both had implanted in him." Max Caspar, *Kepler*, 61. He also notes that Kepler read Proclus "zealously" (92). Kepler mentions Ptolemy repeatedly in his major works.

Ptolemaic Cosmology and Harmonics

Claudius Ptolemy (c. AD 100–170) is best known for his astronomical treatise the *Almagest*, which sets forth a mathematical, geocentric model of the cosmos. In Book I, he draws a distinction between the "theoretical" part of philosophy and what he calls the "practical" part; the former deals with the continually changing stuff of the sublunar region, and the latter is concerned with the unchanging and incorruptible—the "loftiest things of the universe" in the heavenly (supralunar) realm.[2] Between the theoretical and practical lies an intermediate: the mathematical, which can be contemplated in the abstract or perceived in the structure and arrangement of material things:

> And the kind of science which shows up quality with respect to forms and local motions, seeking figure, number, and magnitude, and also place, time, and similar things, would be defined as mathematical. For such an essence falls, as it were, between the other two, not only because it can be conceived both through the senses and without the senses, but also because it is an accident in absolutely all beings both mortal and immortal, changing with those things that ever change, according to their inseparable form, and preserving unchangeable the changelessness of form in things eternal and of ethereal nature.[3]

The mathematical serves as a type of intellectual bridge that enables the natural philosopher to make inquiries into the "beautiful and well-ordered disposition" of nature.[4]

Ptolemy's cosmological system preserved the Aristotelian principle of circular motions in its geometrical descriptions of ob-

---

2 Ptolemy, *The Almagest*, I.1.

3 Ibid.

4 Ibid.

ARISTOTELIAN-PTOLEMAIC COSMOLOGY

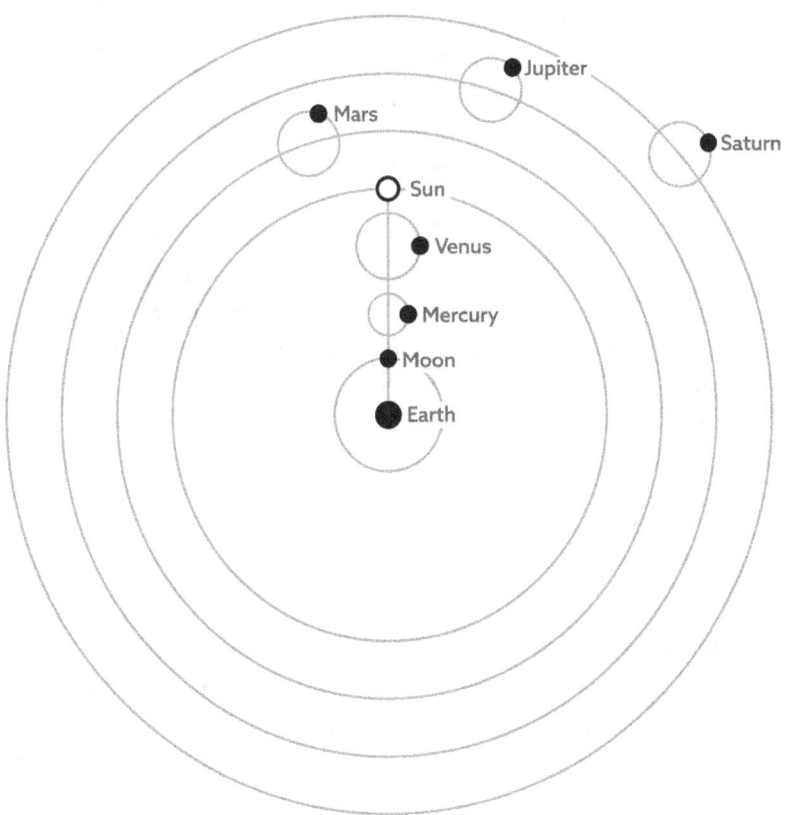

Figure 3.1 *The Aristotelian-Ptolemaic Cosmology*

served irregularities in planetary motion by imposing epicycles upon each body's orbit.[5] Planets were thought to move in comparatively tiny circles as they traveled in one giant circle around the earth, somewhat like the classic Tilt-a-Whirl carnival ride. This helped to "save the phenomena"—to explain why planets sometimes exhibit periods of retrograde motion as they make their way across the sky. Ptolemy's philosophical pre-commitment to perfectly circular paths and uniform speed is expressed in Book IX:

> Now, since our problem is to demonstrate, in the case of the five planets as in the case of the sun and moon, all their apparent irregularities as produced by means of regular and circular motions (for these are proper to the nature of divine things which are strangers to disparities and disorders) the successful accomplishment of this aim as truly belonging to mathematical theory in philosophy is to be considered a great thing, very difficult and as yet unattained in a reasonable way by anyone.[6]

Ptolemy's model reigned supreme for roughly fifteen hundred years before being replaced by heliocentrism, but two underlying philosophical ideas enjoyed continuity in natural philosophy throughout the scientific revolution and beyond: first, the understanding of the cosmos as mathematically intelligible, and second, the application of philosophical principles in mathematical descriptions of the physical world. This is not to say that these ideas were original to Ptolemy, only that they are clearly expressed in his work, which was foundational to astronomy for many centuries.

Another of Ptolemy's works, the *Harmonics*, was also a major influence upon Kepler's cosmological theorizing. Ptolemy criticized certain aspects of Pythagorean thought while preserving the essentials of its harmonic theory, which Kahn describes as "a concern with the physics of sound as well as with the mathematics of

---

5   Ptolemy, *The Almagest*, I.3.

6   Ibid., IX.2.

the octave, and above all, a sense that the whole cosmos is ordered in agreement with the musical numbers."[7] The empirical discovery in harmonics that was attributed to the Pythagoreans is as follows: If a stretched, uniform string (such as that of a monochord) is plucked, resulting in a pitched tone, and then the string is divided in half with a bridge and one of the two sections is plucked, the new tone will be precisely one octave higher than the first. Furthermore, if the original string is divided such that one segment is ⅓ of the original length and the other is ⅔, the longer stretch of string will produce a tone that is a perfect fifth above the pitch of the undivided string. Finally, if the string is divided into ¾ and ¼ lengths, the longer of the two segments will produce a tone that is a perfect fourth above that of the whole string.[8] These intellectually pleasing mathematical relations were seen (by Plato et al.) as the reason for the audible harmony of these tonal pairs.[9] In the *Timaeus*, Plato writes that audible harmonies give pleasure even to the unlearned, but that the learned experience a "higher sort of delight," because the audible harmonies are imitations of the (mathematical) divine harmonies.[10]

The ancient discipline of harmonics was not merely concerned with consonant tones and the associated mathematics. In his preface to his authoritative translation of Ptolemy's *Harmonics*, Jon Solomon defines harmonics in this way: "It is the study of the

---

7   Charles Kahn, *Pythagoras and the Pythagoreans*, 155–156.

8   Andrew Barker, "Ptolemy's Pythagoreans, Archytas, and Plato's conception of mathematics," in *Phronesis* 39, no. 2 (1994): 114.

9   For further discussion see Andrew Barker, "Mathematical Beauty Made Audible: Musical Aesthetics in Ptolemy's *Harmonics*," in *Classical Philology* 110 (2010). Barker cites *Timaeus* 80b when he writes that "the octave is associated with lengths in the ratio 2:1, the fifth with lengths in the ratio 3:2, and the fourth with lengths in the ratio 4:3" (405).

10 Plato, *Timaeus*, 80b. Barker explains that "all audible concords must share some perceptible and aesthetically agreeable attribute that distinguishes them from intervals of all other kinds, and in the same way there must be an attribute of a mathematical sort that is peculiar to the ratios of concords and that justifies the description of such ratios as beautiful" (405).

harmony that can be found in mathematics, music, the human soul, and the cosmos which contains it all. In a word or two, harmonics is the science of cosmic and psychic harmony."[11] This is made clear in Book III, where Ptolemy discusses the connection between the harmonic ratios in the universe and those that exist in the human soul. He explains that mathematical ratios describe the audible differences and "this extends to what becomes the accustomed and proper arrangement created from contemplation and its consequences."[12] In other words, musical harmonies can be described mathematically, and both the human ear and mind are attuned to them. Furthermore, Ptolemy argues, the harmonic ratios can be applied to the zodiac, the fixed stars, and the relationships between the motions of particular planets, the moon, and the sun. Thus, he draws a harmonic connection between mathematics, the physical universe, the sensory experience, and rationality. The heavens exhibit a harmonious construction that can be mathematically described, and the soul of man resonates with these intellectually and aesthetically pleasing harmonies.[13]

Before moving on, it is worth noting that the *Harmonics* cites the same tonal scale used by Plato in the *Timaeus* creation account and incorporates all four of the "sister sciences"—arithmetic, harmonics, geometry, and astronomy—over the course of its description of the harmonious cosmos.[14] For all these reasons, the *Harmonics* is rightfully considered part of the broader Pythagorean-Platonic tradition.

---

11 Ptolemy, *Harmonics*, edited and translated by Jon Solomon (Boston: Brill, 1999), ix.

12 Ibid., 142–143.

13 As will be seen in a later chapter, this idea of a grand cosmic harmony was central to Kepler's natural philosophy and theology.

14 Ptolemy, *Harmonics*, 64 n. 20.

The Pythagorean-Platonic Tradition in the Early Christan Era

Certain Pythagorean-Platonic ideas began to be incorporated into Christian thought as early as the second century. In one of the earliest examples, Pythagorean teaching is specifically mentioned in the *Hortatory Address to the Greeks* written by the apologist Justin Martyr (c. AD 100–165), who refers to Pythagoras's idea of "numbers, with their proportions and harmonies, and the elements composed of both, the first principles" and says that when Pythagoras "says that unity is the first principle of all things, and that it is the cause of all good, he teaches by an allegory that God is one, and alone."[15] Evidence of the integration of Pythagorean-Platonic thought into early theological education comes from Clement of Alexandria (c. AD 150–215).[16] In *The Stromata*, he explains that philosophy is "a preliminary training for the word of the Lord" that is itself preceded by the encyclical studies (which includes the four mathematical arts); just as these arts are handmaidens to philosophy, so philosophy serves the truth and wisdom of God.[17] The learned Christian sifts out truth from all branches of learning "so that, from geometry, and music, and grammar, and philosophy itself" he can retrieve that which is useful and discard what is incompatible with the faith.[18] Against the objection that Christians can be led astray by pagan learning, Clement says that "truth is

---

15 Justin Martyr, *Hortatory Address to the Greeks* in Ante-Nicene Fathers vol. 1, edited by Alexander Roberts and James Donaldson (Peabody, MA: Hendrickson Publishers, 2012), 274, 280.

16 Clement referred to Philo as a Pythagorean rather than a Platonist. See discussion in David Albertson, *Mathematical Theologies: Nicholas of Cusa and the Legacy of Thierry of Chartres* (Oxford: Oxford University Press, 2014), 46.

17 Clement of Alexandria, *The Stromata*, Ante-Nicene Fathers vol. 2, edited by Alexander Roberts and James Donaldson (Peabody, MA: Hendrickson Publishers, 2012), 306.

18 Ibid., 310.

immovable; but false opinion dissolves" and that such studies actually help the faithful distinguish truth from error.[19] Essentially, what the heretics use for wickedness the learned Christian can and should use for righteousness: "He who culls what is useful for the advantage of the catechumens...must not abstain from erudition, like irrational animals; but he must collect as many aids as possible for his hearers" and "be able to take his departure home to the true philosophy, which is a strong cable for the soul, providing security from everything."[20] As will be seen, this is a theme that ran through the works of other prominent Church Fathers as well as Johannes Kepler's.

By the early centuries of Christianity, the "Pythagorean homeland" had long been established in the four mathematical disciplines.[21] Indeed it was, in part, through a defense and use of the Greek *paideia* that Pythagorean-Platonic thought was transmitted into Christian culture.[22] For example, Clement expresses high regard for mathematics, specifically naming each of the four sister sciences in his enumeration of various subjects that have utilitarian and/or philosophical value. Music instructs one on the proportions of harmonies, tempers human emotion and, as the Psalmist demonstrates, can be used for the glorification of God.[23] Arithmetic shows the increasing, decreasing, and relations of numbers, both to each other and to other things.[24] Geometry involves abstract essence that lends much help to architecture and building,

---

19 Clement of Alexandria, *The Stromata*, 499.

20 Ibid., 500.

21 Kahn, *Pythagoras and the Pythagoreans*, 153.

22 Origen (c. AD 185–250) and Gregory Thaumaturgus (AD 213–270) are other excellent examples of this defense. See Origen, *A Letter From Origen to Gregory*, in Ante-Nicene Fathers 4 (393) and Gregory Thaumaturgus, *Oration and Panegyric Addressed to Origen*, in Ante-Nicene Fathers 6 (30).

23 Clement of Alexandria, *The Stromata*, 500.

24 Ibid., 498.

but it also draws the mind to contemplation of the intangible, intelligible realm; it

> ...makes the soul in the highest degree observant, capable of perceiving the true and detecting the false, of discovering correspondences and proportions...conducts us to the discovery of length without breadth, and superficial extent without thickness, and an indivisible point, and transports to intellectual objects from those of sense.[25]

By "intellectual objects," Clement apparently means the abstract geometrical ideas. He goes on to say that astronomy has practical uses such as discerning the seasons and navigation by the position of the stars, but he seems to give special philosophical weight to this discipline. By the study of the heavenly bodies, man is "raised from the earth in his mind, he is elevated along with heaven, and will revolve with its revolution; studying ever divine things...from which Abraham starting, ascended to knowledge of Him who created them."[26] The contemplation of the celestial orbs, the form of the universe, and the movement of the heavens, Clement believes, will lead the soul nearer to knowledge of the Creator. Although he does not mention Plato here, the similarity of the language is striking; yet Clement goes one essential stride further by specifying the Judeo-Christian God as the highest truth and source of all being.

In the third century, as Middle Platonism waned, neo-Pythagorean thought became absorbed into what is now referred to as neo-Platonism, an eclectic reinvented Platonism that began with Plotinus (c. AD 205–270)[27] and was further embellished with

---

25 Ibid., 501.

26 Ibid., 498.

27 Kepler mentions Plotinus several times, especially in his private correspondence. See Rhonda Martens, *Kepler's Philosophy and the New Astronomy* (Princeton: Princeton University Press, 2000) 40. Also, Caspar notes the importance of Plotinus in the intellectual climate of sixteenth-century Tübingen, though he does not elaborate (see Caspar, *Kepler*, 19).

neo-Pythagoreanism by his pupil Porphyry (c. AD 234–305) and Porphyry's pupil Iamblichus (c. AD 245–325).[28] In his *Enneads*, Plotinus conceives of reality as a three-part hierarchy, what is now referred to as the Plotinian hypostases. This hierarchy consists of the One (God)—the ultimate ground of reality from which all else emanates—mind (*nous*), and soul (*psychē*).[29] According to Plotinus, the image of the One becomes weaker at each successive level as it emanates "downward." Emanation from the ineffable One gives rise to *nous* at the intermediary level—*nous* being the hypostasis that contains the Platonic ideas and can be roughly construed as God's intellection. At the third level down is *psychē*, which results from the activity of the *nous* and (among other things) manifests as human souls capable of rational, discursive thought. Plotinus explains that the human soul is, properly speaking, an emanation from the One, even though it is one level removed. In other words, the human soul is not ontologically disconnected from the One, from which it emanates by way of the *nous*: "Sprung, in other words, from the Intellectual-Principle...[and thus] for its perfecting it must look to that Divine Mind, which may be thought of as a father watching over the development of his child born imperfect in comparison with himself."[30]

Plotinus considered the contemplation of the physical world to be a pathway by which the soul may ascend to the *nous*, the realm of the "more authentic" where the archetypes reside:

> Admiring the world of sense as we look out upon its vastness and beauty and the order of its eternal march, thinking of the gods within it, seen and hidden, and the celestial spirits and all

---

28 Kahn, *Pythagoras and the Pythagoreans*, 133ff.

29 This doctrine of emanation is a hallmark of neo-Platonism and was later discussed in the writings of Proclus (AD 410–485) and Nicholas of Cusa (AD 1401–1464), both of whom Kepler read extensively. See Martens, *Kepler's Philosophy and the New Astronomy*, 34.

30 Plotinus, *The Enneads*, Great Books of the Western World 11 (Chicago: Encyclopaedia Britannica, Inc., 1990), 519.

the life of animal and plant, let us mount to its archetype, to the yet more authentic sphere: there we are to contemplate all things as members of the Intellectual...and, presiding over all these, the unsoiled Intelligence and the unapproachable wisdom.[31]

This is a particular point of interest in terms of the ideas being traced in this study; Plotinus sees the sensible realm as something that the human soul, endowed with reason, can contemplate and thereby perceive something of the intellection of the One.

Neo-Platonic Doctrine & Christian Thought

Over the ensuing centuries, facets of neo-Platonic doctrine were integrated into Christian thought through the writings of certain Church Fathers and the work of late medieval theologians. Citing passages such as Psalm 19, these great thinkers affirmed the neo-Platonic idea that the grandeur and ordering of the natural world are a testament to the existence of a transcendent mind with whom mankind has some type of intellectual kinship. A wonderful metaphor emerged to express this revelatory function of the created world—the *liber naturae*, or "book of nature." The exact origin of the metaphor is unclear, but the concept is implied in the writings of at least two of the neo-Platonist Fathers: Athanasius of Alexandria (AD 297–373) and Basil of Caesarea (AD 330–379).[32]

Athanasius often referred to the order and harmony of nature as evidence for God, attributes of the cosmos that are (as the Psalmist wrote) analogous to language. In *Against the Heathen* Atha-

---

31 Ibid., 519–520.

32 Kepler apparently had at least an awareness of the work of Athanasius, as he mentions him among several other Church Fathers in his *Manuscripta Chronologica*. See Kepler's *Gesammelte Werke* 21.2.1, 160. Accessed November 20, 2018 at http://publikationen.badw.de/de/035836814.

nasius writes that it is "possible to attain to the knowledge of God from the things which are seen, since Creation, as though in written characters, declares in a loud voice, by its order and harmony, its own Lord and Creator."[33] Moreover, Athanasius draws a parallel between the Word, or *Logos*, by Whom all things were made (John 1:1–3) and the words of human language. After quoting from the *Book of Wisdom* 13:5—"By the greatness and the beauty of the creatures proportionately the Maker of them is seen"—he writes:

> For just as by looking up to the heaven and seeing its order and the light of the stars, it is possible to infer the Word Who ordered these things, so by beholding the Word of God, one needs must behold also God His Father, proceeding from Whom He is rightly called His Father's Interpreter and Messenger. And this one may see from our own experience; for if when a word proceeds from men we infer that the mind is its source, and by thinking about the word, see with our reason the mind which it reveals, by far greater evidence and incomparably more, seeing the power of the Word, we receive knowledge also of His good Father.[34]

Thus, as we can rightfully ascribe instances of spoken or written language to rational minds, so we can attribute the order and artistic splendor of nature to a higher intelligence—the divine *Logos*. In a stunning passage found in the same work, Athanasius strikes a particularly Pythagorean-Platonic chord:

> [John] the Divine says, "in the beginning was the Word, and the Word was with God, and the Word was God; all things were made by Him and without Him was not anything made." For just as though some musician, having tuned a lyre, and by his art adjusted the high notes to the low, and the intermediate notes to the rest, were to produce a single tune as the result, so also the

---

33 Athanasius, *On the Incarnation* with *Against the Heathen* (Brookline, MA: Paterikon Publications, 2018), 63.

34 Ibid., 77–79.

> Wisdom of God, handling the Universe as a lyre, and adjusting things in the air to things on the earth, and things in the heaven to things in the air, and combining parts into wholes and moving them all by His beck and will, produces well and fittingly, as the result, the unity of the universe and of its order...[35]

Here Athanasius celebrates the grand cosmic harmony in which the observable orderliness of creation is a manifestation of the mind of its source—the divine Word. He says, "For though He be not seen with the eyes, yet from the order and harmony of things contrary it is possible to perceive their Ruler, Arranger, and King."[36]

Like Athanasius, Basil of Caesarea wrote about the purpose of God's natural revelation and the reason human beings are able to perceive it:

> We were made in the image and likeness of our Creator, endowed with intellect and reason, so that our nature was complete and we could know God. In this way, continuously contemplating the beauty of creatures, as if they were letters and words, we could read God's wisdom and providence over all things.[37]

Here, Basil compares the creation to letters and words that we, creatures made in God's image, are equipped to read. Although he does not use the phrase "book of nature," the concept is clearly articulated. It is possible that he was familiar with the metaphor, or that his theological reflections served to inspire it. In any case, a book metaphor was famously used by his contemporary, Augustine of Hippo (AD 354–430), in a passage preserved in the *Sermons*:

> Some read a book to find God. But there is a great book: the spectacle of what has been created. Look upwards and down-

---

35 Ibid., 75.

36 Ibid., 69.

37 Basil of Caesarea, *Homilia de Gratiarum Actione*, as quoted in Alister McGrath, *Christian Theology* (West Sussex: Blackwell Publishing, 2017), 145.

ward; pay attention and read. In order to enable you to read that book, God did not write in letters with ink but he placed what is created itself in front of you. Why do you seek a louder voice? Heaven and earth are crying to you: God made me.[38]

Truly, he argues, could the Creator be any more obvious in revealing Himself? Unlike the text of Scripture, the book of the universe can be read even by the illiterate, so that all men have access to some knowledge of God. Creation conveys the truth to us through our observations and rationality just as words on a page communicate information to a reader. Augustine was actually the originator of the idea that the *liber naturae* and the *liber scripturae* can and should be read in parallel, so that they may illuminate one another.[39] Both the metaphor and the "two books" hermeneutic philosophy went on to flourish throughout the Middle Ages and beyond.[40]

Augustine was clearly influenced by neo-Platonic ideas, notably by the work of Plotinus and Porphyry as well as the Greek neo-Platonist Church Fathers, including Basil and Athanasius.[41] Early in his intellectual career, he was established as a Carthaginian rhetorician, but ambition led him to Italy, where he studied the neo-Platonists and, under the influence of St. Ambrose, converted to Christianity.[42] While planning his return to Africa, he began writing a series of treatises on the liberal arts but unfortunately

---

38 Augustine, *Sermons: 51–94* (Hyde Park, NY: New City Press, 1991), 225–226.

39 Charlotte Methuen, *Kepler's Tübingen* (Brookfield, VT: Ashgate Publishing, 1998), 151.

40 An excellent paper that traces the transmission of the metaphor is Giuseppe Tanzella-Nitti, "The Two Books Prior to the Scientific Revolution," *Perspectives on Science and Christian Faith* 57, no. 3 (September 2005): 235–248.

41 John J. O'Meara, "The Neoplatonism of Saint Augustine," in *Neoplatonism and Christian Thought*, ed. Dominic J. O'Meara (Norfolk: International Society for Neoplatonic Studies, 1982), 35, 39.

42 Mortimer Adler, "Biographical Note: Saint Augustine," Great Books of the Western World 16 (Chicago: Encyclopaedia Britannica, Inc., 1990), v.

On Music is the only one that has survived.[43] His general attitude towards pagan learning is that it should be used in service to Christianity, that we should "claim it for our own use from those who have unlawful possession of it."[44] He writes: "For, as the Egyptians had not only the idols and heavy burdens which the people of Israel hated and fled from, but also vessels and ornaments of gold and silver, and garments, which the same people when going out of Egypt appropriated to themselves, designing them for a better use, not doing this on their own authority, but by the command of God."[45] In referencing the liberal arts, Augustine stresses the importance of knowledge (symbolized by the golden and silver vessels) that can help in understanding theological truths but pointedly rebukes the pridefulness that can result from higher learning, especially when the Author of all wisdom is not credited or served by it. In his *Confessions*, he laments his own mistakes in this regard: "I understood all that I read on the arts of rhetoric and logic, on geometry, music, and mathematics...But since I made no offering of them to you, it did me more harm than good..."[46] In *On Christian Doctrine*, he argues against what he deems superfluous learning—that which does not contribute to the Church—but makes a noteworthy concession to the restrained study of mathematics, the "sciences of reasoning and of number."[47]

In his discussion of the utility of the various arts, Augustine prefaces his opinion on mathematics with commentary on its origin: "Coming now to the science of numbers it is clear to the dullest apprehension that this was not created by man, but was discov-

---

43 Ibid., v.

44 Augustine, *On Christian Doctrine*, Great Books of the Western World vol. 16 (Chicago: Encyclopaedia Britannica, Inc., 1990), 737.

45 Ibid..

46 Augustine, *The Confessions*, Great Books of the Western World 16 (Chicago: Encyclopaedia Britannica, Inc., 1990), 33.

47 Augustine, *On Christian Doctrine*, 737.

ered by investigation...[numbers] have fixed laws which were not made by man, but which the acuteness of ingenious men brought to light."[48] In other words, the truths of mathematics are *discovered* rather than *invented*. He supports his argument by pointing out that it is in no man's power to determine the attributes and relationships of numbers, which are demonstrated by arithmetical operations. The philosophical implication is that these abstract entities point beyond man's intellect and the physical world they describe to a transcendent reality.

Augustine significantly tempered the Pythagorean-Platonic tradition before handing it on to medieval Christianity.[49] Number symbolism is important in Augustine's view, but he believes it should be limited to numbers mentioned in Scripture.[50] "Ignorance of numbers...prevents us from understanding things that are set down in Scripture in a figurative and mystical way," he explains.[51] He goes on to give examples of the symbolic use of number, which he regards as a legitimate practice in theology:

> ...the number ten signifies the knowledge of the Creator and the creature, for there is a trinity in the Creator; and the number seven indicates the creature, because of the life and the body. For the life consists of three parts, whence also God is to be loved with the whole heart, the whole soul, and the whole mind; and it is very clear that in the body there are four elements of which it is made up.[52]

---

48 Augustine, *On Christian Doctrine*, 736. Augustine goes on to say that prideful man, boasting that he is "one of the learned" fails to "inquire after the source from which those things which he perceives to be true derive their truth." His point seems to be that what the learned men know ultimately depends upon God for its objective truth.

49 See discussion in David Albertson, *Mathematical Theologies: Nicholas of Cusa and the Legacy of Thierry of Chartres*, 69.

50 Augustine, *On Christian Doctrine*, 737.

51 Ibid., 726.

52 Ibid.

Augustine believes that without a knowledge of number symbolism, certain meanings of Scripture remain obscure, thus the science of numbers "is found to be of eminent service to the careful interpreter."[53] He also applies number symbolism to music theory, emphasizing once again, in a clearly Pythagorean-Platonic fashion, the importance of the number ten. It is perfectly reasonable, he says, "for learned men to discuss, whether there is any musical law that compels the psaltery of ten chords to have just so many strings; or whether, if there be no such law, the number itself is not on that very account the more to be considered as of sacred significance."[54]

Augustine shows less regard for astronomy than the other mathematical disciplines, perhaps because of its misuse by astrologers, the so-called "diviners of the fates."[55] He perceives virtually no profit in the knowledge of the stars and their courses where interpretation of Scripture is concerned, and considers intensive astronomical study a distraction from worthwhile occupations.[56] He excludes from criticism the common understanding of the lunar phases, which are "employed in reference to celebrating the anniversary of our Lord's passion."[57] In *The Confessions*, he criticizes the highly skilled astronomers who have had enormous success in correctly predicting solar and lunar eclipses but have become arrogant in their own abilities:

> They lapse into pride without respect for you, my God...they ignore you and do not inquire how they come to possess the intelligence to make these researches. Even when they discover that it was you who made them, they do not submit to you...They do

---

53 Augustine, *The City of God*, Great Books of the Western World 16 (Chicago: Encyclopaedia Britannica, Inc., 1990), 393.

54 Augustine, *On Christian Doctrine*, 727.

55 Ibid., 733.

56 Ibid.

57 Ibid.

not know Christ, who is the Way and the Word of God, by which you created all the things which they number and count, the very men who count them, the senses by which they are aware of what they count, and the intelligence by which they count them.[58]

The folly of the astronomers is the same one that Augustine confesses committing—the failure to dedicate all knowledge to God and give credit to Him for all of creation, including the rationality of mankind that makes study of the orderly world possible in the first place.

In his commentary on Genesis, Augustine sounds much like *Wisdom of Solomon* and Philo when he specifically mentions measure, number, and weight "in which, as it is written, God has arranged all things."[59] Like Philo, Augustine was intent upon characterizing the Platonic Forms—the immaterial patterns for creation—in a way that made them compatible with Christian monotheism:

> ...who would dare to say that God has created all things without a rational plan?...individual things are created in accord with reasons unique to them. As for these reasons, they must be thought to exist nowhere but in the very mind of the Creator. For it would be sacrilegious to suppose that he was looking at something placed outside himself when he created in accord with it what he did create.[60]

Once again, the underlying plans for the cosmos are seen as ideas in the Creator's mind rather than external, independently existing abstractions. Augustine insists that there was nothing outside of God before the original act of creation, and therefore these things existed "in him" as ideas.[61] As for humanity's apprehension of the

---

58   Augustine, *The Confessions*, 35.

59   Augustine, *On Genesis* (New York: New City Press, 2002), 246.

60   Augustine, *Eighty-three Different Questions* in *The Fathers of the Church*, vol. 70 (Washington, DC: CUA Press, 2010), 80–81.

61   Ibid., 248.

orderliness of creation, Augustine explains that the soul of man is illumined by God in a way that enables us to detect the rationality that pervades the world.[62] It is because we are made in the image of the Creator Himself that we are cognizant of His natural revelation to us.

The Transmission of Neo-Platonic Thought Through Pagan Writings

It is important to understand that some pagan writers also played a key role in the transmission of neo-Platonism to the Middle Ages. Proclus Diadochus (AD 412–485), whose work had an important influence on Kepler's natural philosophy, was regarded as one of the best interpreters of Plato's dialogues until later modern efforts to purify Plato of the "taint of neo-Platonism."[63] He is best known for his commentaries on ancient works, which included (among many others) several Platonic dialogues, Euclid's *Elements*, Nicomachus's *Introduction to Arithmetic*, as well as a (now lost) commentary on the *Enneads* of Plotinus. Proclus drew upon the work of his neo-Platonic predecessors, including Plotinus's student and editor, Porphyry, with whom he shared an intellectual opposition to Christianity.[64] Proclus saw the writings of the neo-Platonic philosophers as the reemergence of a faithful understanding of Platonic doctrine.[65]

---

62 Ibid., 81.

63 Danielle Layne, ed., *Proclus and His Legacy* (Boston: De Gruyter, 2017), 1. Layne writes, "Indeed it was only late into the modern age that Proclus' stature as an invaluable exegete of Plato...became overshadowed...by those who wished to return to a kind of Platonism 'purified' from the so-called taint of neo-Platonism." No approximate date range is given.

64 Proclus, *A Commentary on the first Book of Euclid's Elements*, trans. Glenn R. Morrow (Princeton: Princeton University Press, 1970), xiv-xv.

65 Ibid., xvi.

The work that is relevant to the present discussion is Proclus's commentary on Euclid's *Elements*, on account of the fact that Kepler studied it extensively.[66] One of Proclus's chief goals in the commentary was to rectify what he perceived as a lack of Platonic (actually neo-Platonic) context in the *Elements*.[67] He lauds the importance of mathematics, saying that "mathematics makes ready our understanding and our mental vision for turning towards that upper world...For the beauty and order of mathematical discourse, and the abiding and steadfast character of this science, bring us into contact with the intelligible world itself."[68] Mathematics is mental preparation for training what Proclus calls the "eye of the soul" towards the world of Platonic Forms; it releases the soul from the cave so that it may ascend to reality.[69] About the utility of mathematics in natural philosophy (translated here as "physical science") Proclus writes, "Mathematics...makes contributions of the very greatest value to physical science. It reveals the orderliness of the ratios according to which the universe is constructed and the proportion that binds things together in the cosmos, making, as the *Timaeus* somewhere says, divergent and warring factors into friends and sympathetic companions."[70] Notice that Proclus refers to the harmonic ratios believed to be fundamental to the orderly structure of the world as well as the idea of the harmonizing of unlike things into one coherent cosmos.

---

66 Caspar, *Kepler*, 92. There is direct evidence that Kepler also read Proclus's commentary on the *Timaeus*, but he cites the commentary on Euclid far more often, particularly in the openings to Book I and Book IV of the *Harmonice Mundi*.

67 Guy Classens, "Imagination as Self-knowledge: Kepler on Proclus' Commentary on the First Book of Euclid's Elements" in *Early Science and Medicine* 16 (2011): 185.

68 Proclus, *A Commentary on the first Book of Euclid's Elements*, 17.

69 Ibid., 17, 23–24.

70 Ibid., 19. Kepler quotes this passage and others in the opening of Chapter IV of the *Harmonice Mundi*.

Proclus explains that mathematical demonstrations direct the mind of the student to "Nous, from which it gets its principles."[71] Here he alludes to the neo-Platonic metaphysical hypostatic hierarchy (the One, *nous*, and *psychē*), in which the human soul is derivative of *psychē*, which is in turn an emanation from the One. Like Ptolemy, Proclus sees mathematical ideas as intermediate, existing in a middle ground between immaterial realities (Forms) and the sensible world of divisible things; he believes that education in mathematics aids in man's ascent from ordinary existence towards the higher plane of being—*nous*, the abstract realm of the Forms.[72] Proclus rejected the Aristotelian idea that mathematical ideas are intellectually extracted from sensible material objects. He says,

> But they cannot come from sense objects, for then there would be far more precision in sense objects than there is. They come therefore from the soul, which adds perfection to the imperfect sensibles and accuracy to their impreciseness...Do we not see that all sensible things are confused with one another and that no quality in them is pure and free of its opposite, but that all are divisible and extended and changing?[73]

Proclus understands geometrical figures in particular as perfect, immaterial archetypes and sees sensible matter as their receptacle:

> But if the objects of geometry are outside matter, its ideas pure and separate from sense objects, then none of them will have any parts or body or magnitude. For ideas can have magnitude, bulk, and extension in general only through the matter which is their receptacle, a receptacle that accommodates indivisibles as divisible, unextended things as extended, and motionless things as moving.[74]

---

71   Proclus, *A Commentary on the first Book of Euclid's Elements*, 23.

72   Ibid., xix, 3.

73   Ibid., 11.

74   Ibid., 40.

In other words, the sensible world is the material manifestation of the abstract objects of geometry. This archetypal conception of the world and the soul's access to it was, as aforementioned, central to Kepler's metaphysics.

CHAPTER 4

The Pythagorean-Platonic Tradition in the Middle Ages

Two translators were especially important to the transmission of the Pythagorean-Platonic tradition from late antiquity into the Middle Ages: Calcidius, who is believed to have lived during the late fourth century, and Boethius (c. AD 480–524).[1] Nearly nothing is known about the life of Calcidius. Possibly a Christian, he translated Plato's *Timaeus* from Greek to Latin, and it was this translation, along with his own commentary, that survived and came to be identified with the Platonism of the Middle Ages.[2] As Andrew Hicks puts it, Calcidius's work was "the primary conduit through which Plato (and Platonic Pythagoreanism) reached the Latin West."[3] This translation and commentary is by no means an uncolored lens; as C. S. Lewis has critically noted, Calcidius's work "is hardly what we should call a commentary, for it ignores many difficulties and expatiates freely on matters about which Plato had little or nothing to say."[4] Of particular interest is the fact that Calcidius believed that the rational and harmonic

---

1 David Lindberg, *The Beginnings of Western Science* (Chicago: University of Chicago Press, 2007), 147.

2 Ibid.

3 Andrew Hicks, "Pythagoras and Pythagoreanism in late antiquity and the Middle Ages," in Carl Huffman, *A History of Pythagoreanism* (Cambridge: Cambridge University Press, 2014), 428-429.

4 C. S. Lewis, *The Discarded Image* (New York: Cambridge University Press, 2012), 49.

structure of the world (which is resonant with the Platonic World Soul[5]) is accessible to the human soul by way of mathematics.[6] He writes, "Because soul was designed to penetrate both surfaces and solids with its vital vigor, it was necessary that it possess powers akin to the solid [i.e., the cubic numbers 9 and 27] and the surface [i.e., the square numbers 4 and 8], insofar as like flocks with like."[7] By "surfaces" he presumably means two-dimensional objects, and by "solids" three-dimensional objects. The phrase "like flocks with like" seems to indicate that there is a kinship between the rational basis of material reality (which is modeled after an immaterial rationality) and the mathematical powers of the human soul.

Boethius, a Christian neo-Platonist philosopher, translated Euclid's *Elements* as well as Nicomachus's *Introduction to Arithmetic* and *Handbook of Harmonics*.[8] These three works became integral to the mathematics curriculum known henceforth as the *quadrivium* (arithmetic, geometry, harmonics, and astronomy).[9] Boethius's use of this term, which is Latin for "the place where four roads meet," is the earliest in extant literature.[10] In his *De institutione arithmetica* (c. AD 515), which preserves Nicomachus's *Introduction to Arithmetic*, he writes about the importance of the *quadrivium*:

---

5   The World Soul, or *anima mundi*, in Plato's *Timaeus* is the rational, animating substance that pervades all of the cosmos.

6   Andrew Hicks, "Pythagoras and Pythagoreanism in late antiquity and the Middle Ages," 430.

7   Calcidius, *On the Timaeus*, quoted in Hicks, "Pythagoras and Pythagoreanism in late antiquity and the Middle Ages," 430. Here, Hicks has erroneously switched 9 (a square number) with 8 (a cubic number).

8   Charles Kahn, *Pythagoras and the Pythagoreans*, 110–111.

9   As has been shown, the idea of four mathematical "sister sciences" goes back at least as far as Plato.

10   P. G. Walsh, "Introduction," in Boethius, *Consolation of Philosophy* (New York: Oxford University Press, 2008), xxi.

> Among all the ancient men of authority who, following the lead of Pythagoras, have flourished in the purer reasoning of the mind, it is clearly obvious that hardly anyone has been able to reach the highest perfection of the disciplines of philosophy unless the nobility of such wisdom was investigated by him in a certain four-part study, the quadrivium, which will not be hidden to a just and penetrating mind. For this is the wisdom of things that are, and the perception of truth gives to these things their unchanging character.[11]

Boethius referred to the *quadrivium* as the "fourfold road that must be traversed by those whom a more excellent soul leads away from the senses inborn within us to the greater certainties of understanding."[12] The concept of the *quadrivium* was nothing new, of course, but the Christian liberal arts tradition was responsible for its perpetuation. As David Albertson explains, "The history of Pythagoreanism is the prehistory of the *quadrivium*; the *quadrivium's* history unfolded within the long, pluriform doctrinal and pedagogical ambit of medieval Christianity."[13]

Boethius's work suggests that the reconciliation of Christianity with elements of the Pythagorean-Platonic tradition (as it was understood at the time) was regarded as feasible. In his famous *Consolation of Philosophy*, he speaks of Lady Philosophy repeatedly reminding him of the "dictum of Pythagoras: 'Follow the god.'"[14] More than once he refers to the beauty and order of the heavens and credits the transcendent Creator:

---

11 David Albertson, *Mathematical Theologies: Nicholas of Cusa and the Legacy of Thierry of Chartres*, 82.

12 Quoted in Andrew Hicks, "Pythagoras and Pythagoreanism in late antiquity and the Middle Ages," in Carl Huffman, *A History of Pythagoreanism*, 422.

13 David Albertson, *Mathematical Theologies*, 10.

14 Boethius, *Consolation of Philosophy*, 12.

> Creator of the starry sphere,
> Seated upon your timeless chair,
> You move the sky in swift gyration,
> Ordering with law each constellation.[15]

At one point, Lady Philosophy asks, "Do you think that the course of the world is random and haphazard, or do you believe it is guided by reason?" and Boethius replies, "I should certainly refuse to believe…that such unerring movements are the outcome of random chance. I know that the creator God superintends his creation. The day will never come when I detach myself from the truth of that statement."[16] In his work on harmonics, he discusses "cosmic music," which he describes as the periods of the celestial orbs and their harmonic structures.[17]

The late Middle Ages ushered in a marked revival of interest in mathematical study, and mathematics was considered the essential key that unlocks the workings of the world.[18] This, of course, involved the Pythagorean-Platonic tradition in its later form. Andrew Hicks explains that

> medieval Pythagoreanism arises from a reification and simultaneous iconization of a particular strain of late ancient (Neo-)Pythagorean speculation. For it was the "mathematical Pythagoreanism" of Nicomachus (and, indirectly, Iamblichus and Proclus…) that most captured the medieval imagination…it was this Pythagoras and "Pythagorean" speculation that was subsequently inherited and enriched by Renaissance philosophers and humanists.[19]

---

15 Boethius, *Consolation of Philosophy*, 13.

16 Ibid., 16.

17 Hicks, "Pythagoras and Pythagoreanism in late antiquity and the Middle Ages," 422.

18 David Lindberg, *The Beginnings of Western Science*, 42.

19 Hicks, "Pythagoras and Pythagoreanism," 419.

Thus, it was the neo-Pythagorean understanding of what constituted "Pythagoreanism" that went on to be held by subsequent thinkers, including theologians and the great natural philosophers of the scientific revolution. "It was not a question of what Pythagoras 'really' thought," explains Hicks, "but rather a question of how to deploy appropriately the Pythagorean inheritance within the new cultural and philosophical contexts of (largely) Christian patterns of thought."[20]

By the twelfth century, the *Timaeus* had become central to the project of natural philosophy, and it effectively propagated neo-Pythagorean mathematical speculation about all things having a fundamental relationship to number.[21] Scholars placed great emphasis upon the mathematical orderliness of nature, which was wrought by a divine craftsman and included human beings and their intellectual powers.[22] "Having made humankind part of the natural order," historian David Lindberg writes, "twelfth-century scholars increasingly took an interest in the 'natural man' and his capacities"; because it was "part of the natural order, and therefore sympathetic to its rhythms and harmonies, reason was regarded as a particularly suitable instrument for the exploration of the cosmos."[23] As previously discussed, this is one of the ancient ideas that would go on to become a central tenet of Kepler's natural philosophy.

The twelfth and thirteenth centuries produced newly translated Greek works, including those of Aristotle and Ptolemy, which paved the way for their geocentric cosmology to be the ruling paradigm in the centuries preceding the scientific revolution.[24] John

---

20 Ibid., 419.

21 Lindberg, *The Beginnings of Western Science*, 209.

22 Ibid.

23 Ibid., 214.

24 Ibid.

of Salisbury (c. AD 1120–1180), Bishop of Chartres, provides another example of how Christian philosophers were interacting with the Pythagorean-Platonic tradition in their expositions on the created order and the nature of man. In his *Metalogicon*, John insists that if the Platonic Forms exist at all, they must have come from "him by Whom all things have been made."[25] He sounds much like Philo, *Wisdom of Solomon*, and Augustine when he says that "God is number innumerable, weight incalculable, and measure inestimable. And in Him alone all things that have been made in number, in 'weight,' and in measure, have been created."[26] Concerning the rational powers that separate men from brutes he says, "For God, breathing life into man, willed that he partake of the divine reason."[27] Here again we see the major themes of Kepler's natural theology: a rationally ordered world and God's image-bearers as the only creatures mentally equipped to contemplate its implications. Further on, John makes a statement that is quite reminiscent of Plato: "Reason...transcends all sense perception, and judges concerning spiritual as well as material realities. Not only does it consider all things found here [on earth] below, but it also rises to the contemplation of heavenly things."[28]

One of John's early scholastic contemporaries, Thierry of Chartres (AD 1100–1150), was instrumental in the continuation of the Pythagorean-Platonic tradition in both philosophical and theological thought.[29] He wrote that

---

25 John of Salisbury, *The Metalogicon* (Philadelphia: Paul Dry Books, 2009), 29.

26 Ibid., 133.

27 Ibid., 227.

28 Ibid., 228.

29 Thierry became chancellor of the cathedral school at Chartres in 1141. This was roughly a half century before the construction of the Chartres Cathedral that still stands today. The cathedral's structure and adornments contain extensive neo-Pythagorean symbolism. One of Thierry's students, Hermann of Carinthia, called Thierry the reincarnation of Plato. See David Albertson, *Mathematical Theologies*, 95.

none of the ancients who are regarded as great undertook difficult questions by any other than mathematical likenesses. Thus, Boethius, the most learned of the Romans, maintained that without some training in mathematics no one could attain a knowledge of divine things. Did not Pythagoras, the first philosopher in name and deed, locate every investigation of truth in the study of numbers? The Platonists and our own major thinkers have followed him to such a degree that our Augustine, and later Boethius, declared that of the things to be created number was undoubtedly "the principle exemplar in the mind of the Creator."[30]

As Thierry's words attest, numbers continued to be seen as an archetype for the created order—ideas that pre-existed in the mind of God. In terms of applying mathematics to nature, natural philosophers often discussed matters such as the sphericity of the earth, its circumference, and the motions of the celestial bodies.[31] Thierry wrote a commentary in which he related Platonic cosmology to the Genesis creation narrative: *Treatise on the Work of the Six Days*. There he quotes Genesis 1:2—which speaks of God moving over the waters of the formless void—and comments, "Having described matter, [Moses] then says that the power of the artificer, which he calls the spirit of God, presided over matter and ruled it in order to inform and order it...Plato, in the *Timaeus*, calls the same spirit the 'world soul.'"[32] Thierry was also largely responsible for a renewed emphasis on the Boethian *quadrivium*;[33] he says that the four mathematical disciplines lead to knowledge of the exis-

---

30 Albertson, *Mathematical Theologies*, 177–178.

31 Lindberg, *Western Science*, 254.

32 Thierry of Chartres, *Treatise on the Work of the Six Days*, Katharine Park, trans., 11. Accessed November 3, 2022 at https://www.academia.edu/31388090/Thierry_of_Chartres-Treatise_ Six_Days-trans._Park.pdf. This association between the Jewish hexaemeron and the *Timaeus* would later be made by Kepler, who thought of the latter as a commentary on the former. This will be briefly discussed in Chapter 6.

33 Albertson, *Mathematical Theologies*, 94.

tence of a rational creator: "There are four kinds of arguments that lead man to knowledge of the Creator, namely the proofs of arithmetic, music, geometry, and astronomy. These tools should be used briefly in this theology so as to make rationally apparent—as I propose—the artifice of the Creator of things."[34]

Thierry's legacy passed on to Nicholas of Cusa (AD 1401–1464), a German Catholic cardinal whose work significantly shaped Kepler's natural philosophy.[35] "There is no other premodern author," writes Albertson, "who so fully took on the enterprise of a Christian Neopythagoreanism as Nicholas of Cusa."[36] During an ecumenical stint in Constantinople in the 1430's, Nicholas uncovered manuscripts of works by neo-Platonists such as Basil of Caesarea and Proclus, and those by Proclus apparently included some of his commentaries on Plato.[37] Nicholas's subsequent work, *On Learned Ignorance*, contains some high points of his philosophy, such as the idea that the mathematical disciplines are guides to the study of creation, as well as other clear references to Pythagorean-Platonic doctrine.[38] He writes:

> All our wisest and most divine teachers agree that visible things are truly images of invisible things and that from created things

---

34 Thierry of Chartres, *Treatise on the Work of the Six Days*, 11.

35 Charlotte Methuen, citing R. S. Westman, writes that Kepler's understanding of the Christian-Platonic tradition came "from Pythagoras through Augustine to Cusanus" (Methuen, *Kepler's Tübingen*, 21). Caspar says that it is established that Kepler had read "various writings by Nicholas of Cusa, whose geometrical mysticism agreed so closely with his own thinking" (Caspar, *Kepler*, 44). However, it is not certain that Kepler had direct access to *On Learned Ignorance*.

36 Albertson, *Mathematical Theologies*, 10.

37 Ibid., 170. A fascinating fact about Nicholas of Cusa is that he established (1458) a hospice home for elderly men, the Cusanus-Stift or St. Nikolaus Hospital, which he endowed both financially and with his extensive personal library. This library, which includes over three hundred volumes, includes manuscripts written in Nicholas's own hand. The Cusanus-Stift still functions as originally intended and offers tours of the chapel and library.

38 Ibid., 171.

> the Creator can be knowably seen as in a mirror and a symbolism...for all things have a certain comparative relation to one another, (nonetheless, hidden from us and incomprehensible to us), so that from out of all things there arises one universe and in one maximum all things are this one...Therefore, in mathematicals the wise wisely sought illustrations of things that were to be searched out by the intellect. And none of the ancients who are esteemed as great approached difficult matters by any other likeness than mathematics. Thus, Boethius, the most learned of the Romans, affirmed that anyone who altogether lacked skill in mathematics could not attain a knowledge of divine matters. Did not Pythagoras, the first philosopher both in name and in fact, consider all investigation of truth to be by means of numbers? The Platonists and also our leading [thinkers] followed him to such an extent that our Augustine, and after him Boethius, affirmed that, assuredly, in the mind of the Creator number was the principal exemplar of the things to be created.[39]

It is evident that Nicholas saw the created world as mathematically ordered according to archetypes ("exemplars") in the mind of God, and thus intelligible to mankind by way of the mathematical sciences.

Nicholas often reiterates his belief that the Creator God (the Maximum, Absolute Being, and Necessity) is the one ultimate self-existent reality, the source of all other being.[40] He asks, "For how could that which is not from itself exist in any other way than from Eternal Being?"[41] In other words, if something is, by nature, not self-existent, it must have an eternal, self-existent source. He says that "all things are either the Absolute Maximum or from

---

39 Nicholas of Cusa, *On Learned Ignorance*, in *Complete Philosophical and Theological Treatises of Nicholas of Cusa: Volume One*, translated and edited by Jasper Hopkins (Minneapolis: Banning Press, 1990), 18–19.

40 Ibid., 61.

41 Ibid., 61–62.

the Absolute Maximum."[42] This includes the Platonic Forms—the immaterial rational patterns for the created world—which depend upon the Infinite Form (God) for their existence.[43] God is the "Form of forms" and the *ratio* of all things.[44] Created things (everything outside of God) are contingent and reflect God's perfection only in an approximate way, due to their finitude.[45] Nicholas directly criticizes the Platonists for their belief that the Forms are a multitude of ideas with independent existence, arguing that "it is not possible that there be many maximal and most true things. For only one infinite Exemplar is sufficient and necessary; in it all things exist."[46] The Platonists, he says, conceive of a "World Soul" that is the "connecting necessity" between the "Begetting Mind" and the manifestation of intellect in the material world, but he believes that this is incorrect.[47] "Rather," he says, "it is the divine Word and Son, equal with the Father. And it is called 'Logos' or 'Essence,' since it is the Essence of all things."[48] It is the divine *Logos*, the second person of the Trinity through whom all things were created, as holy Scripture testifies in John's gospel: "all form and actuality exist in Absolute Form, which is the Father's divine Word and Son, so all uniting motion and all uniting proportion and harmony exist in the Divine Spirit's Absolute Union, so that God is the one Beginning of all things. In Him and through Him all things exist in a certain oneness of trinity."[49]

---

42 Nicholas of Cusa, *On Learned Ignorance*, 68.

43 Ibid., 64.

44 Ibid., 77.

45 Ibid., 64.

46 Ibid., 85. Here, Nicholas simply refers to "the Platonists." He apparently meant early Platonists, because he mentions Aristotle's attempt to refute them on some of their teachings.

47 Ibid., 86.

48 Ibid.

49 Ibid., 89. This theme will be discussed at length in Chapters 9 and 10.

Nicholas goes on to describe his cosmology, in which the earth is *not* the "fixed and immovable center" of the "world-machine," and the fixed stars are not the outer circumference of the created world, as the Aristotelian-Ptolemaic system requires.[50] This is a notable pre-Copernican instance of a non-geocentric understanding of the universe (he denied that it was possible for the universe to even *have* a center, which was an even more radical idea than heliocentrism) and an anticipation of the mechanistic philosophy that flowered during the scientific revolution. Nicholas writes that God Himself is, metaphorically speaking, "the [only] center and circumference" of the cosmos.[51] He explains that a terrestrial observer cannot depend upon the apparent motionlessness of the earth and movement of the heavenly bodies to deduce a stationary earth around which the rest of the universe revolves, since the appearance of motion is always relative to the observer's vantage point.[52] He asks, "For example, if someone did not know that a body of water was flowing and did not see the shore while he was on a ship in the middle of the water, how would he recognize that the ship was being moved?"[53] He argues that it is the same with any earth-bound observer of the heavens.

From the created order, Nicholas writes, we reach some understanding of God: "It is the unanimous opinion of the wise that visible things—in particular, the size, beauty, and order of things—lead us to an admiration for the divine art and the divine excellence."[54] He lists the sciences of the *quadrivium* to explain the ways in which we study God's mathematical ordering of the world:

---

50 Ibid., 90.

51 Ibid.

52 Ibid., 91.

53 Ibid., 93. The idea of being limited by one's reference frame is often attributed to Galileo, but here we see that the idea is much older. Moreover, Kepler explored it in his *Somnium*.

54 Ibid., 98.

In creating the world, God used arithmetic, geometry, music, and likewise astronomy. (We ourselves also use these arts when we investigate the comparative relationships of objects, of elements, and of motions.) For through arithmetic God united things. Through geometry He shaped them, in order that they would thereby attain firmness, stability, and mobility in accordance with their conditions. Through music He proportioned things...And so, God, who created all things in number, weight, and measure, arranged the elements in admirable order. (Number pertains to arithmetic, weight to music, measure to geometry.)[55]

In his commentary on the relationship of number to the cosmos and the *quadrivium* as the set of tools man uses to understand it, he echoes Philo, *Wisdom*, Augustine, and John of Salisbury with his mention of "number, weight, and measure." He goes on to laud the Artisan, Who arranged the heavenly bodies with "such skill that there is—though without complete precision—both a harmony of all things and a diversity of all things...He established the interrelationship of parts so proportionally that in each thing the motion of the parts is oriented toward the whole."[56] The world is a finely-tuned system, and God wills that mankind know Him, the "Truth of all things" through "admiring so marvelous a world-machine."[57]

The Capstone: Nicolaus Copernicus

This survey now concludes with one of Kepler's most important predecessors, Nicolaus Copernicus (AD 1473–1543), the Polish astronomer who was on his deathbed when his treatise on heliocentrism, *On the Revolutions of the Heavenly Spheres*, was first pub-

---

55   Nicholas of Cusa, *On Learned Ignorance*, 99.
56   Ibid., 100.
57   Ibid., 101.

lished.[58] During his lifetime, the Ptolemaic system was still reigning supreme; the common claim that due to accumulating astronomical observations it had become increasingly complicated to the point of near-collapse is historical myth.[59] Yet, Copernicus saw the Ptolemaic system as a pieced-together geometric monstrosity that contradicted "the first principles of regularity of movement" and did not reveal "the form of the world and the certain commensurability of its parts."[60] He was convinced that the heavens must be arranged with much greater symmetry and harmony, as they (as the Psalmist writes) reflect the wisdom and glory of God.

Copernicus devised a heliocentric system for calculating the past and future positions of the heavenly bodies, but despite its comparative conceptual elegance, it was not substantially superior to Ptolemy's system in its predictive accuracy.[61] Moreover, it was not the product of any new empirical data. As Thomas Kuhn explains, "No fundamental astronomical discovery, no new sort of astronomical observation, persuaded Copernicus of ancient astronomy's inadequacy or of the necessity of change."[62] Copernicus seems to have preferred heliocentrism, at least in part, for theological and aesthetic reasons; as Owen Gingerich explains, "Although observational evidence could not have entered directly

---

58   Heliocentrism was not a new idea, having been postulated at least as far back as Aristarchus of Samos in the third century BC.

59   Owen Gingerich, *Copernicus* (New York: Oxford University Press, 2016), 33. An example of how this falsehood has even crept into respected academic literature is a passage in Alister McGrath's textbook, *Science & Religion: A New Introduction* (West Sussex: Blackwell Publishers, Ltd., 2010): "Initially, the discrepancies could be accommodated by adding additional epicycles. By the end of the fifteenth century, the model was so complex and unwieldy that it was close to collapse" (18).

60   Nicolaus Copernicus, *Revolutions of the Heavenly Spheres*, Great Books of the Western World vol. 15 (Chicago: Encyclopaedia Britannica, Inc., 1990), 507.

61   Owen Gingerich, *The Eye of Heaven: Ptolemy, Copernicus, Kepler* (New York: American Institute of Physics, 1993), 171.

62   Thomas Kuhn, *The Copernican Revolution* (Cambridge: Harvard University Press, 1995), 132.

into Copernicus's enthusiasm for the heliocentric layout, aesthetic considerations undoubtedly played a powerful role."[63] Copernicus indeed found his system "pleasing to the mind" in various ways. For example, it situated the fastest moving planet, Mercury, in the innermost orbit, and the slowest, Saturn, in the outermost, with the other planets falling into a gradation in between the two. He remarked that such an arrangement could be "found in no other way."[64] In the tenth chapter of Book I, he expresses the appropriateness of the sun being the center of all creation:

> In the center of all rests the sun. For who would place this lamp of a very beautiful temple in another or better place than this where from it can illuminate everything at the same time? As a matter of fact, not unhappily do some call it the lantern; others, the mind and still others, the pilot of the world…And so the sun, as if resting on a kingly throne, governs the family of stars which wheel around.[65]

He goes on to say that this arrangement of the world "has a wonderful commensurability" and there is "a sure bond of harmony for the movement and magnitude of the orbital circles such as cannot be found in any other way."[66] The use of the word "harmony" should not be overlooked; as Aviva Rothman has explained, "Because the Copernican cosmos was unified and symmetrical in ways that the Ptolemaic was not, it was heralded by its promulgators as closer to the true Pythagorean vision."[67] Kuhn underscores the fact that Copernicus's method of persuasion was "to point out how

---

63 Owen Gingerich, "The Copernican Revolution," in Gary Ferngren, ed., *Science & Religion: A Historical Introduction* (Baltimore: Johns Hopkins University Press, 2017), 89.

64 Ibid.

65 Copernicus, *Revolutions of the Heavenly Spheres*, 526–528.

66 Ibid., 528.

67 Aviva Rothman, *The Pursuit of Harmony: Kepler on Cosmos, Confession, and Community* (Chicago: University of Chicago Press, 2017), 20.

The Pythagorean-Platonic Tradition in the Middle Ages

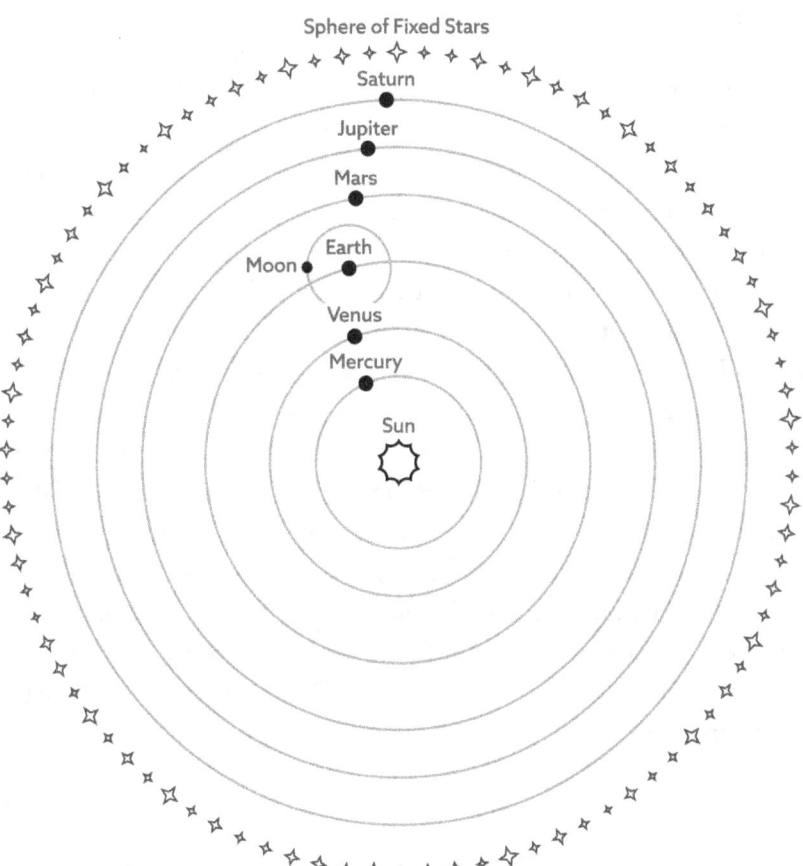

Figure 4.1 *The Copernican Cosmology*

much more harmonious, coherent, and natural" his model was.[68] Copernicus's arguments were not pragmatic; as Kuhn explains, "They appeal, if at all, not to the utilitarian sense of the practicing astronomer but to his aesthetic sense and to that alone...They did not necessarily appeal to astronomers, for the harmonies to which Copernicus's arguments pointed did not enable the astronomer to perform his job better. New harmonies did not increase accuracy or simplicity."[69] However, Kuhn adds, they did appeal to the subgroup of mathematical astronomers who possessed a "neo-Platonic ear for mathematical harmonies."[70] The work of these few astronomers (most notably Kepler), turned out to be an essential element of the scientific revolution.

As a side note, the Copernican love of harmony is further attested by Copernicus's sole pupil, Georg Joachim Rheticus. In his *Narratio prima* (AD 1540), Rheticus writes that the Copernican cosmos "imitate[s] the musicians who, when one string has either tightened or loosened, with great care and skill regulate and adjust the tone of all the other strings, until all together produce the desired harmony and no dissonance is heard in any."[71] He says that in the heliocentric view, "the entire harmony of the celestial motions is established and preserved" under the control of the sun.[72] This Copernican idea that the sun functions to govern the motions of the planets is a stunning anticipation of later developments in natural philosophy.[73]

---

68 Thomas Kuhn, *The Copernican Revolution*, 181.

69 Ibid.

70 Ibid.

71 Peter Pesic, *Music and the Making of Modern Science* (Cambridge: MIT Press, 2014), 47.

72 Ibid.

73 First, Kepler's postulation of the *anima motrix*, a force emanating from the sun and driving the planets, and later, Isaac Newton's elucidation of gravitational force. Both will be discussed in later chapters.

Copernicus's preference for elegant order and harmony is apparent in his prefatory dedication to Pope Paul III, where he writes that the system of the universe is framed *for the sake of mankind* by "the Best and Most Orderly Workman of all" Who has so precisely arranged the spheres of the heavens that "nothing can be shifted around in any part of them without disrupting the remaining parts and the universe as a whole."[74] Later, in a discussion of the great distance between Saturn and the sphere of fixed stars, he again expresses aesthetic appreciation for the details of his model: "It is by this mark in particular that they [the fixed stars] are distinguished from the planets, as it is proper to have the greatest difference between the moved and the unmoved. How exceedingly fine is the godlike work of the Best and Greatest Artist!"[75] Copernicus's use of the adjective "godlike" is notable in that it strongly hints towards the concept of God's mind being made manifest in nature and assumes that man's aesthetic sensibilities are in tune with the Creator's.

Like many others before him, Copernicus regards the mathematical art of astronomy as a study that raises the mind upward to contemplation of God. He writes that astronomy is "the head of all the liberal arts and the one most worthy of a free man" and asks "who, after applying himself to things which he sees established in the best order and directed by divine ruling, would not through diligent contemplation of them and through a certain habituation be awakened to that which is best and would not wonder at the Artificer of all things…?"[76] For Copernicus, God was the Artist and Architect of the heavens, the One Who planned and formed the world in attunement with beautiful mathematical harmonies.

---

74 Copernicus, *Revolutions of the Heavenly Spheres*, 508.

75 Ibid., 529.

76 Ibid., 510.

Thanks to the work of Copernicus, the Pythagorean-Platonic tradition, Christianity, and astronomy became integrated in such a way that the stage was well set for Johannes Kepler's grand entrance.

CHAPTER 5

The Early Formation of Kepler's Natural Philosophy

With the relevant intellectual history complete, we now turn to Johannes Kepler's development as a natural philosopher. An overview of Kepler's formal education at Tübingen will connect the dots from the foregoing intellectual history to his studies and mentorship at the university. This will be followed by an analysis of the relevant philosophical and theological ideas made manifest in his earlier major works: *Mysterium Cosmographicum*, a book he published as a young mathematics instructor at Graz, and the *Astronomia Nova*, which is considered his most valuable contribution to the rise of modern science.

Kepler's University Years

In 1589, with his sights set on becoming a theologian, Johannes Kepler began his studies at the University of Tübingen, a Lutheran institution with a curriculum largely fashioned by Philip Melanchthon, an intellectual leader in the German Reformation.[1] Melanchthon saw nature as a mode of God's revelation and there-

---

1 Rhonda Martens, *Kepler's Philosophy and the New Astronomy* (Princeton: Princeton University Press, 2000), 11.

fore promoted natural philosophy as an integral part of theological training.² Although he championed all seven liberal arts in his educational program, he particularly prized arithmetic, geometry, and astronomy because of his conviction that reality is fundamentally mathematical and thereby reflects divine rationality.³ Moreover, he viewed the human aptitude for mathematics as the part of the *imago Dei* that gives mankind access to God through the study of nature. He understood the neo-Pythagorean notion of mind as number to mean that the human soul is a rational being that comprehends mathematical order and thus requires training in the mathematical arts in preparation for the study of philosophy.⁴

Astronomy was an integral component of Melanchthon's curriculum, as it was understood as a way to apprehend God's natural revelation to mankind. Although he considered it the most important of the mathematical sciences, arithmetic and geometry were the "wings" that raised the human mind to the contemplation of the celestial sphere.⁵ Melanchthon believed that "vestiges of divinity" can be detected in the heavens, traces of the mind of God in the regularities of the celestial motions.⁶ Historian Charlotte Methuen elaborates:

> Melanchthon explains that Paul in his letter to the Romans (1.20) encourages the study of philosophy in order that 'God's presence in nature' can be considered, for 'the whole of the universe is a

---

   2   Charlotte Methuen, *Kepler's Tübingen*, 71. Note that Methuen's work (book and articles) constitutes the definitive scholarly treatment of this topic.

   3   Ibid., 12. For more on Melanchthon's emphasis on the mathematical arts, see Endnote 7.

   4   Charlotte Methuen, "The Role of the Heavens in the Thought of Philip Melanchthon," *Journal of the History of Ideas* 57, no. 3 (July 1996): 390.

   5   Methuen, *Kepler's Tübingen*, 75.

   6   Ibid., 82. The concept of "vestiges of divinity" was not original to Melanchthon; it can be found in the work of earlier theologians, such as the thirteenth-century Franciscan, Bonaventure. See Endnote 8.

sort of sacrament, because it is a testimony that God is, and that God is wise, good, just'. He points out that the human mind has been formed by God to study the heavens and to recognize the vestiges of divinity, but reminds his readers that the Word of God rules this philosophy, just as it does everything else. There is no mention of any special status for astronomy, although the movements of the heavenly bodies are specifically cited as evidence for God's existence.[7]

Melanchthon's idea here is that the creation contains sensible manifestations of God's mind that are accessible to the human mind on account of the latter's mathematical aptitude. Methuen explains that, according to Melanchthon, "the human mind may be said to be number in its capacity to seek out order and regularity, and…in this it reflects the mind of God, and in this way human observation of the heavens is able to offer a route to better knowledge of God."[8]

As a result of Melanchthon's substantial influence upon the curriculum at Tübingen, Kepler was required to spend his first two years hearing lectures by the faculty of arts, which included mathematics, astronomy, and physics.[9] The program could rightfully be described as Christian humanist, as it included both Latin and Greek and taught the seven liberal arts as an integrated body of knowledge secondary to theology. Melanchthon's ideas were actively discussed among the faculty at that time, and his pupil, Jacob Heerbrand, taught both Kepler and Michael Maestlin, Kepler's most influential professor.[10] Heerbrand's writing is permeated with affirmations of natural theology—specifically the belief that God may be perceived in the structure and organization of nature.[11] The

---

7   Ibid., 87.

8   Ibid., 83.

9   Max Caspar, *Kepler*, 42.

10 Methuen, *Kepler's Tübingen*, 102, 108–109. There is, however, no evidence that actual texts written by Melanchthon were used in any of the courses.

11 Ibid., 109.

influence of the Pythagorean-Platonic tradition is evident in his work. For example, in the introduction to his lectures on Genesis, he speaks of various creation accounts, including Plato's *Timaeus*.[12]

During his years of study at the university, Kepler was undoubtedly exposed to the Pythagorean-Platonic tradition, though precisely how and to what extent is, unfortunately, not entirely clear. As Max Caspar explains, Kepler says little about the specific works that were included in his studies at Tübingen, but it is evident that "from the very beginning his whole thinking was stamped in accordance with Platonic and Neo-Platonic speculation" and that "from the system of ideas which tradition connects with the name of Pythagoras, he received the strongest impetus for his work."[13] This was, no doubt, largely the result of Melanchthon's educational program as well as the transmission of his ideas through professors such as Heerbrand and Maestlin. It is known with certainty, however, that Kepler read the works of various neo-Platonists and had a particular affinity for the writings of Nicholas of Cusa (who may have been Kepler's introduction to Proclus).[14]

Maestlin, a professor of mathematics and astronomy and a highly esteemed astronomer of the time, became one of Kepler's most important and longest-lasting scholarly influences. Maestlin's instruction at Tübingen included Euclid's *Elements* as well as his own textbook, *Epitome Astronomia*.[15] He inspired Kepler's adoption of the heliocentric Copernican system, which Maestlin saw as mathematically superior to the reigning geocentric Ptolemaic system officially taught at the university (Maestlin's text, however, was based upon the Ptolemaic system).[16] Kepler came to embrace

---

12 Methuen, *Kepler's Tübingen*,, 113.

13 Caspar, *Kepler*, 44.

14 Ibid.

15 Ibid., 46.

16 Ibid. For further discussion on Maestlin's teaching versus his personal convictions, see Endnote 9.

Copernicanism without having read Copernicus's full treatise, *On the Revolutions of the Heavenly Spheres*, for himself; it was Maestlin's arguments that originally persuaded him of heliocentrism's superiority.[17] Kepler explains:

> ...when I was studying under the distinguished Master Michael Maestlin at Tübingen, I was disturbed by the many difficulties of the usual conception of the universe, and I was so delighted by Copernicus, whom Mr. Maestlin often mentioned in his lectures, that I not only frequently defended his opinions at the disputation of candidates in physics but even wrote out a thorough disputation on the first motion, arguing that it comes about by the Earth's revolution...I collected together little by little, partly from the words of Maestlin, partly by my own efforts, the advantages which Copernicus has mathematically over Ptolemy.[18]

Thus, even as a young university student, Kepler was participating in scholarly debate in which he overtly defended the merits of Copernicanism. He was already developing strong intellectual preferences concerning the mathematical structure of the cosmos.

Despite his fascination with—and talent for—mathematics and astronomy, Kepler did not abandon his original goal of becoming a professional theologian. As Caspar explains, "Kepler had not gone to Tübingen to become a philosopher, a mathematician, or an astronomer. Everything which he absorbed in the faculty of arts was only supposed to serve as preparation for the theological studies which opened the gate to the desired church office."[19] How-

---

17   Ibid., 47. However, historian of science Owen Gingerich explains that Kepler studied at least one section of Copernicus's book alongside Maestlin at Tübingen in the early/mid 1590s and owned a secondhand copy by the time he went to Prague to work with Tycho Brahe in 1600. See Owen Gingerich, *The Book Nobody Read: Chasing the Revolutions of Nicolaus Copernicus* (New York: Walker, 2004), 163–165. Caspar notes that Kepler had a copy in his possession at least by 1598. See Caspar, *Kepler*, 47 n. 1.

18   Johannes Kepler, *Mysterium Cosmographicum* (Norwalk, CT: Opal Publishing, 1981), 63.

19   Caspar, *Kepler*, 48.

ever, during his time at Tübingen, Kepler realized that he did not entirely agree with the Formula of Concord, a statement of faith that all Lutheran clergy were required to affirm.[20] This presented a double obstacle for Kepler: he was not well-suited for a career in the church *or* in academia, since university appointments were only open to clerics.[21] Fortunately, he was presented with a viable alternative: he was invited to Graz to fill the position of mathematics instructor at the Protestant *Stiftsschule*, and took up this post in April of 1594. The encouragement Kepler received from his professors to accept the position was not an attempt on their part to rid themselves of a Copernican: "No," says Caspar, "Kepler was sent to Graz because, on the basis of his mathematical and astronomical knowledge, he was by far the most suitable candidate for the teaching position there, the only one worthy of consideration and likely to bring honor to Tübingen University."[22] Neither was it on account of any theological dispute; in fact, it was not until years later that Kepler made known his theological leanings (which were dissonant with the Formula of Concord).[23]

The *Mysterium Cosmographicum*

During his time at Graz, Kepler taught mathematics, astronomy, rhetoric, Virgil, history, and ethics, and was regarded as a first-rate scholar but not a talented teacher. Caspar elaborates:

---

20   Peter Barker and Bernard R. Goldstein, "Theological Foundations of Kepler's Astronomy," *Osiris* 16 (2001): 96. The Formula of Concord (1577) was an interpretation of the Augsburg Confession that elaborated the Lutheran position on various theological controversies. Kepler rejected the Formula's statements on the ubiquity of Christ's body and the doctrine of predestination.

21   Ibid.

22   Caspar, *Kepler*, 52.

23   Ibid.

> It seldom happens that a great scholar, rich in ideas, or a creative genius is at the same time a good teacher. This applies also to Kepler. If he found few listeners, then the fault was certainly in part his. He expected too much of his pupils and assumed they would have the same intellectual flexibility and receptiveness, the same enthusiasm for his subject and the same devotion to the knowledge of truth by which he himself was animated.[24]

When he was not confined to the classroom, Kepler zealously studied writings from the Pythagorean-Platonic tradition and was particularly drawn to mathematical texts such as Proclus's commentary on the first book of Euclid's *Elements*.[25] It was during this phase of his career that Kepler's emerging cosmology—fully shaped by his liberal arts education and infused with Pythagorean-Platonic philosophy—began to develop on the trajectory that would eventually lead to his renown in natural philosophy. "Consciously or unconsciously," writes Caspar, "Kepler's thoughts were connected with everything which he had heard and read of Pythagoras and Plato, of Augustine and Nicholas of Cusa and many other great men of the past and with that which Christian teaching about God and the world and the position of men regarding both had implanted in him."[26]

Kepler was captivated by certain questions related to the (Copernican) structure of the universe, such as the reason for the number of planets and the distances between their orbits, and why planetary speed varies based upon proximity to the sun.[27] He wondered if the key to these mysteries was to be found in aesthetically

---

24 Ibid., 57.

25 Ibid., 92.

26 Ibid., 61.

27 Kepler, *Mysterium Cosmographicum*, 63.

pleasing geometrical constructions.[28] Thus, in his aptly named first book, *Mysterium Cosmographicum (The Secret of the Universe)*, published in 1597, Kepler proposed a geometrical model in which each of the so-called Platonic solids—the five regular polyhedra (cube, tetrahedron, octahedron, dodecahedron, and icosahedron)—were circumscribed by nested spheres, each of which represented one of the planetary orbits. These were not the nested, solid crystalline spheres of the Aristotelian-Ptolemaic model, a concept that Kepler regarded as "absurd and monstrous"; rather, they were conceptual spheres that merely served as three-dimensional representation of the planetary orbits.[29] Kepler describes the arrangement as follows:

> The Earth is the circle which is the measure of all. Construct a dodecahedron round it. The circle surrounding that will be Mars. Round Mars construct a tetrahedron. The circle surrounding that will be Saturn. Now construct an icosahedron inside the Earth. The circle inscribed within that will be Venus. Inside Venus inscribe an octahedron. The circle inscribed within that will be Mercury.[30]

In this geometrical scheme, the sizes of the spheres relative to one another indicates the relative circumferences of the planetary orbits.

Kepler was convinced that he had, in a divine flash of insight, discovered the explanation for the existence of exactly six planets (Neptune and Uranus had not yet been discovered) and their respective orbital circumferences.[31] He writes:

---

28   This Pythagorean-Platonic view of the relationship between geometry and nature shaped Kepler's thinking about many other aspects of human life. An excellent, book-length treatment of this fact is Aviva Rothman's *The Pursuit of Harmony: Kepler on Cosmos, Confession, and Community*.

29   Kepler, *Mysterium Cosmographicum*, 167.

30   Ibid., 69.

31   For Kepler's detailed account of the classroom demonstration that sparked his geometrical epiphany (which he dates to July of 1595), see *Mysterium Cosmographicum*, 65–67.

The Early Formation of Kepler's Natural Philosophy

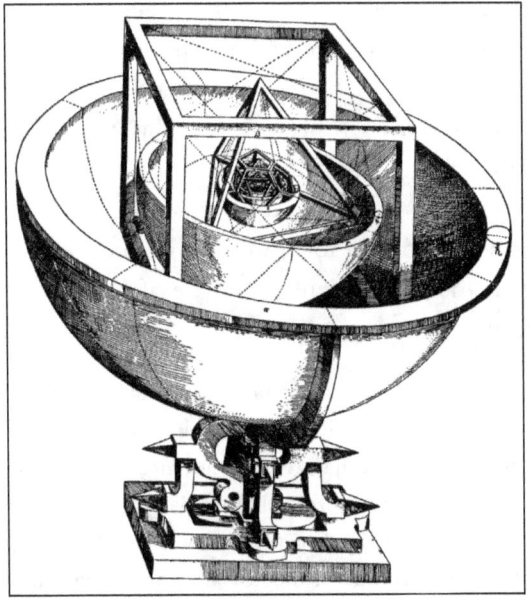

Figure 5.1 *Kepler's diagram of the Polyhedral Model from* Mysterium Cosmographicum

Eventually by a certain mere accident I chanced to come closer to the actual state of affairs. I thought it was by divine intervention that I gained fortuitously what I was never able to obtain by any amount of toil…the happy ending of my toil…What delight I have found in this discovery I shall never be able to express in words…I made a vow to Almighty God that at the first opportunity I would proclaim among men in public print this wonderful example of his wisdom.[32]

Kepler believed that his polyhedral theory revealed the archetype of creation, an archetype of mathematical elegance grounded in the mind of a Creator God Who desired a rationally and visually beautiful world. For Kepler, there was no question that God's con-

---

32 Ibid., 65, 69.

cept of beauty bears similarities to man's, since man is made in the image of God and thus has an intellectual kinship with Him. That the number of regular polyhedra was exactly what was needed to account for the number of planets (known at the time) thrilled Kepler because of his firm conviction about the existence of a pleasing mathematical plan behind the cosmos.

On October 3, 1595, a few months after conceiving his polyhedral theory as the pre-existent archetype of the universe, Kepler wrote to Maestlin,

> I am eager to publish soon, not in my interest, dear teacher...I strive to publish them in God's honor who wishes to be recognized from the book of nature. But the more others continue in these endeavors, the more I shall rejoice; I am not envious of anybody. This I pledged to God, this is my decision. I had the intention of becoming a theologian. For a long time I was restless: but now see how God is, by my endeavors, also glorified in astronomy.[33]

Any residual uncertainties and unfulfilled longings Kepler may have harbored after taking up a mathematics position instead of clerical robes seem to have evaporated with his newfound conviction that he could instead use his God-given abilities in mathematics to exegete the book of nature.

Kepler's goal in the *Mysterium* was to prove, both physically and metaphysically, that his model reflected the Creator's archetypal plan, for he was convinced that "God, like one of our own architects, approached the task of constructing the universe with order and pattern, and laid out the individual parts accordingly, as if it were not art which imitated Nature, but God Himself had looked to the mode of building of Man who was to be."[34] Caspar elaborates:

---

33 Carola Baumgardt, *Johannes Kepler: Life and Letters* (New York: Philosophical Library, 1951), 31.

34 Kepler, *Mysterium Cosmographicum*, 54–55.

> [Kepler] sought the answers to his questions in geometry, in the structure of space. For the geometrical figures have their foundation in the divine being, and consequently it is in them that one should seek the numbers and sizes which appear in the visible world. Everything is regulated according to mass and number. The world is established in accordance with the norms of the quantities provided by geometry. That was why God also gave people a mind which is able to recognize these patterns.[35]

Kepler believed there to be a special resonance between the mind of man and the mind of God regarding mathematical order and harmony and that this could serve as an *a priori* guide in the investigation of nature. As Bruce Stephenson explains, "Kepler believed...that his own sense of harmony reflected the Creator's preferences. It represented, therefore, an order that was objectively present in the universe."[36] For Kepler, intellectual beauty in descriptions of nature were more likely to reflect truths about the structure of physical reality.

The ideas that pervade Kepler's *Mysterium Cosmographicum* are undoubtedly the result of Pythagorean-Platonic influence. For example, in the *Timaeus*, Plato employed four of the regular polyhedra as the basis for the geometrical foundation of the material world. The Pythagorean character in the dialogue, Timaeus, describes the sequential construction of the tetrahedron (pyramid), octahedron, icosahedron, and the cube, and assigns each of these to one of the four elements that were believed to be, in some sense, fundamental to the material world: fire, air, water, and earth, respectively. Aiton, Duncan, and Field, in the introduction to their authoritative English translation of Kepler's *Harmonice Mundi*, highlight the fact that although Kepler rejected Plato's theory of geometrical ele-

---

35 Caspar, *Kepler*, 62.

36 Bruce Stephenson, *The Music of the Heavens: Kepler's Harmonic Astronomy*, 96.

ments, there is a striking similarity between the Platonic cosmology and the spirit of Kepler's polyhedral model:

> For Kepler, the regular polyhedra or Platonic solids provided a key to the structure of the planetary system. In developing this application, he used mathematics in a way that was very similar to that of Plato in the *Timaeus*, but whereas Plato only produced a vague qualitative theory, Kepler succeeded in devising a testable mathematical model of the cosmos.[37]

Thus, Kepler seems to have been inspired not only by the intellectual aesthetic of the regular polyhedra, but also by the manner in which Plato regarded geometry as a formal cause for the material world. Kepler situates his cosmology firmly within his theological framework; as one scholar puts it,

> Kepler's cosmology—Christian to the core and written two thousand years after Plato—closely relates to the cosmogony outlined in Plato's *Timaeus*...Like Plato, he conceives of the cosmos as a copy of the divine model, and again like Plato, he believes that this copy came about by geometric ordering. Indeed, Kepler sees geometry as the archetype of the cosmos, coeternal with the Creator himself, and therefore preceding the Creation of Heaven and earth.[38]

Although Kepler was well studied in Euclid, who gives a proof for the set of five regular polyhedra in Book XIII of the *Elements*, the preface to the *Mysterium* specifically mentions Pythagoras and Plato:

> It is my intention, reader, to show in this little book that the most great and good Creator, in the creation of this moving universe,

---

37 E. J. Aiton, A. M. Duncan, and J. V. Field, "Introduction," in Johannes Kepler, *The Harmony of the World* (Philadelphia: American Philosophical Society, 1997), xiv.

38 Albert van der Schoot, "Kepler's search for form and proportion," *Renaissance Studies* 15, no. 1 (March 2001): 60.

and the arrangement of the heavens, looked to those five regular solids, which have been so celebrated from the time of Pythagoras and Plato down to our own, and that he fitted to the nature of those solids, the number of the heavens, their proportions, and the law of their motions.[39]

Later, in Chapter II, Kepler writes, "What else remains except to say with Plato, 'God is always a geometer'" and then goes to on explain that "in this structure of moving stars he [God] has inscribed solids within spheres, and spheres within solids, until no further solid was left which was not robed outside and inside with moving spheres."[40] Although Kepler does not specifically link his polyhedral model with the *Timaeus* in this passage, it is significant that he mentions Plato's alleged remark about geometry in that exact context. Charles Kahn affirms the strong connection between the Platonic and Keplerian cosmologies:

> Kepler explicitly recognizes the *Timaeus* as his model...Just as Plato's demiurge builds the world out of number series, elementary triangles, and regular solids, so Kepler's God is also a consummate geometer, whose plan for the world can be penetrated by the human mind only if it succeeds in discovering the mathematical relationships realized in celestial phenomena.[41]

Moreover, in the same chapter of the *Mysterium*, Kepler muses about what God's reason might be for giving the universe the physical structure it possesses. His conclusion on the matter indirectly references the *Timaeus*: "'For it neither is nor was right' (as Cicero

---

39   Kepler, *Mysterium Cosmographicum*, 63.

40   Ibid., 97. The phrase "God always geometrizes" was attributed to Plato by the historian Plutarch. See Yuval Se'eman, "Plato Alleges that God Forever Geometrizes," *Foundations of Physics* 26, no. 5 (1996): 575.

41   Charles Kahn, *Pythagoras and the Pythagoreans: A Brief History*, 162.

in his book on the universe quotes from Plato's *Timaeus*) 'that he who is the best should make anything except the most beautiful.'"[42]

In the summer of 1597 (just a few months after Kepler's marriage to a young widow named Barbara Müller), Galileo Galilei received a copy of the *Mysterium*. He immediately read the introduction and then penned an enthusiastic letter of appreciation to Kepler in which he calls Kepler "a companion in the exploration of truth" and laments that "it is deplorable that there are so few who seek the truth."[43] Here, Galileo presumably refers to the academics and churchmen who refused to abandon the heliocentric Aristotelian model and consider the merits of Copernicanism. Kepler's response, written some weeks later, implored Galileo to go public with his preference for heliocentrism and expressed philosophical agreement with him by noting that together, they were "following the lead of Plato and Pythagoras, our true masters."[44]

The *Astronomia Nova*

In early August of 1600, Kepler was officially banished from Graz by the Archduke for refusing to convert to Catholicism. Deprived of his teaching salary and facing a precarious financial situation, Kepler's saving grace took the form of an invitation to become the assistant to Tycho Brahe, renowned astronomer and Imperial Mathematician of Prague. Upon his departure from Graz in September, Kepler wrote the following words to his beloved mentor and erstwhile professor: "I would not have thought that it is so sweet, in companionship with some brothers, to suffer injury and indignity for the sake of religion, to abandon house, fields,

---

42 Kepler, *Mysterium Cosmographicum*, 93. Here, Kepler quotes from the *Timaeus*, 30a–30b.

43 Carola Baumgardt, *Johannes Kepler: Life and Letters*, 38–39.

44 Ibid., 41.

friends and homeland. If it is this way with real martyrdom and with the surrender of life and the exultation is so much the greater, the greater the loss, then it is an easy matter also to die for faith."[45] Along with his wife and stepdaughter, Kepler journeyed towards Prague, where he would spend the twelve most productive years of his career.

Kepler's early months in Prague were not without their challenges and frustrations. The 54-year-old Brahe closely guarded his immense accumulation of astronomical records, and shared only those that were essential to Kepler's assignments. This somewhat strained collaboration lasted less than a year; Brahe died unexpectedly of what was likely a severe urinary tract infection on October 24, 1601, merely ten months after the installation of his new assistant. As Brahe's successor, Kepler inherited both an imperial position and a treasure trove of extensive and precise stellar observations. Equipped with the best empirical data of his day, he launched into an investigation of celestial physics that was published several years later under the title *Astronomia Nova* (1609), a text now regarded as one of the classics of early modern science.[46]

The *Astronomia Nova* was revolutionary; as Owen Gingerich remarks, "Kepler's work was truly the 'new astronomy'…it was the introduction of physics into astronomy that was Kepler's most fundamental contribution."[47] Indeed, Kepler's thinking was far ahead of most of his contemporaries in that he sought both a mathematically and physically rigorous model of the known uni-

---

45 Letter to Michael Maestlin dated September 9, 1600, in *Johannes Kepler Gesammelte Werke* 14, ed. Max Caspar (Munich: Verlag, 1955), 175.

46 This is reflected in the full-length title of his book: *New Astronomy Based upon Causes, or Celestial Physics Treated by Means of Commentaries on the Motions of the Star Mars, from the Observations of Tycho Brahe.*

47 Owen Gingerich, "Kepler and the Laws of Nature," *Perspectives on Science and Christian Faith* 63, no. 1 (March 2011): 17.

verse.⁴⁸ In other words, he was not content with a mere geometrical description of the planetary motions; he desired to describe the causes in operation. Notably, several years after publication of the *Astronomia Nova*, Kepler was chided by Maestlin for this approach to natural philosophy. In a letter dated September 1616, Maestlin argues that geometry and arithmetic are the proper foundations of astronomy, and that speculating on physical causes is more confusing than informative.⁴⁹ Yet, it was Kepler's ground-breaking work that served as the springboard for the subsequent rise of modern astronomy. Kepler had concluded that the great animating principle that drove the six planets (including the earth) in their orbits was an immaterial magnetic force emanating from the rotating sun.⁵⁰ The closer a planet came to the sun, which was situated off-center within the planet's orbital path, the stronger the effect of this force and the faster the planet moved (in other words, with greater angular velocity). In this line of thinking, Kepler was groping his way towards the conception of gravity that Newton (armed with a proper understanding of inertia) later elucidated.

The *Astronomia Nova* contains the first two of Kepler's famous three planetary laws. The so-called Second Law says that a planet sweeps out an equal area in an equal amount of time when a line segment is imagined joining a planet to the sun (a curious fact, since the angular velocity varies based upon distance from the sun). When Kepler began calculating the orbit of Mars, he realized that this "area rule" would not hold unless the orbit was elliptical (rath-

---

48   Jack J. Lissauer, "In Retrospect: Kepler's *Astronomia Nova*," *Nature* 462 (December 2009): 725.

49   Michael Maestlin, "Maestlin to Kepler, 21 September 1616," *Johannes Kepler Gesammelte Werke* 17, ed. Max Caspar (Munich: Verlag, 1955), 187. Writes Maestlin, "Calculus enim fundamenta Astronomica ex Geometria et Arithmetica, suis videlicet alis, postulat, non coniecturas physicas, quae lectorem magis perturbant, quam informant" ("For the reasoning of Geometry and Arithmetic provide the foundation for Astronomy, not physical conjectures, which more disturb than inform the reader.")

50   Kepler explains this idea in Chapters XXXIII and XXXIV of the *Astronomia Nova*.

er than perfectly circular) with the sun at one focus of the ellipse.[51] The claim that planetary orbits are elliptical came to be known as the First Law.[52] As will be seen, these two laws (as well as the Third Law, which Kepler described years later) would be mathematical conclusions of Newtonian mechanics, though Newton did not give Kepler due credit.[53] Gingerich notes that the *Astronomia Nova* is "the first published account wherein a scientist documents how he has coped with the multiplicity of imperfect data to forge a theory of surpassing accuracy."[54] The introduction to the treatise, a passage which had been salvaged from the material censored out of the first edition of the *Mysterium*, was the only significant piece of Kepler's work translated into English and widely read by non-specialists.[55] It will be revisited in a later chapter.

---

51   Owen Gingerich, "Kepler and the Laws of Nature," 19.

52   The mathematics associated with the ellipse had been explored more than a millennium before Kepler's time. For example, see Apollonius of Perga's *On Conic Sections*, which was included in the first edition set of Britannica's Great Books of the Western World, but was later excluded.

53   In a letter Isaac Newton wrote to Edmund Halley dated June 20, 1686, Newton said that Kepler had merely guessed that the planetary orbits were elliptical whereas he himself had proven it.

54   Owen Gingerich, Foreword to Johannes Kepler, *Astronomia Nova*, trans. William Donahue (Santa Fe, NM: Green Lion Press, 2015), xii.

55   Ibid., xiii.

KEPLER'S 1st LAW
THE LAW OF ELLIPSES

All planets orbit the Sun in elliptical orbits with the Sun as one common focus.

KEPLER'S 2nd LAW
THE LAW OF EQUAL AREAS

The line between a planet and the Sun (the radius vector) sweeps out equal areas in equal time.

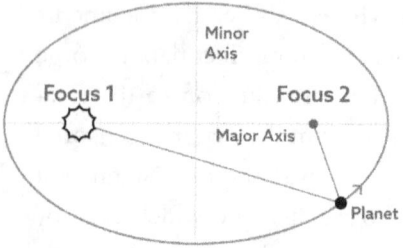

Elliptical orbit of a planet
(greatly exaggerated)

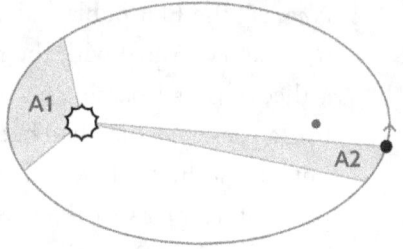

A1 (area 1) = A2 (area 2)

KEPLER'S 3rd LAW
THE LAW OF PERIODS

The square of a planet's period, P, is directly proportional to the cube of its mean distance from the Sun, a.

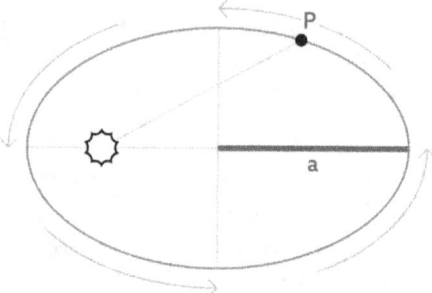

P = orbital period
a = mean distance from sun

Figure 5.2 *Kepler's Three Planetary Laws*

Chapter 6

Kepler's Magnum Opus: The Harmonice Mundi

The year 1611 brought great sorrow and hardship, ending a comparatively blessed period of Kepler's life. The unfortunate events actually began at the end of 1610, when his wife endured a severe case of Hungarian fever coupled with epileptic episodes and began showing signs of mental illness. She experienced a measure of recovery by January, but then the three Kepler children contracted smallpox. Friedrich, the six-year-old son greatly beloved by the family, died in February. Meanwhile, political developments brought bloody warfare to their doorstep. Kepler's imperial master, Holy Roman Emperor Rudolf II, was forced to abdicate in May; Prague was collapsing. Only weeks before, Kepler had sought and failed (on account of his theological reservations with the Formula of Concord) to obtain a professorship back in his homeland of Württemberg. He was considered for a professorship in Padua that had been vacated by Galileo, but despite a recommendation from Galileo, the post was awarded to someone else. In June, he signed a contract to serve as district mathematician in Linz, a post that he expected would greatly please his wife and prevent financial troubles. When he arrived home with the good news, he discovered that his wife was extremely ill again, this time with spotted typhus, a highly contagious infection thought to have been brought to Prague by Austrian troops. She died on the third of July.[1]

At Rudolf's request, Kepler remained in Prague for several months after the loss of his wife. During that time, he published a work entitled *Eclogae Chronicae*, which dealt with the chronology of the life of Christ. He finally departed for Linz in April of 1612 (not long after Rudolf's death) and took up his new residential post in May. His tenure as district mathematician would last fourteen years, but those years were not free of tumult. Only weeks after his arrival, he was denied communion by the Lutheran church because of his theological position about the Formula of Concord, which was painful, but he could not, in good conscience, defy what he thought of as the "burden of antiquity"—the theological testimony of early patristic literature.[2] Just a few years after Kepler's arrival in Linz, the Counter Reformation escalated in Austria, further complicating his religious situation. Then, in 1620 and 1621, his service as the legal defense in his mother's witch trial demanded a great deal of his time and mental energy.[3] Considering the difficulties he faced during these years at Linz, it is no small wonder that Kepler managed to produce what he considered his magnum opus: the *Harmonice Mundi* (1619).

It should be re-emphasized at this juncture that harmony—the "fitting together" of disparate parts to make a coherent and aesthetically pleasing whole—was a central metaphysical principle of Kepler's natural philosophy. His fervent pursuit of a mathematically harmonious cosmos is already evident in his *Mysterium*, where he uses one geometrical device (the nested Platonic solids)

---

[1] The biographical details here described are drawn from Max Caspar's book, *Kepler*.

[2] Letter to Hafenreffer dated April 11, 1619, in *Johannes Kepler Gesammelte Werke* 17, ed. Max Caspar (Munich: Verlag, 1955), 835.

[3] Kepler's defense of his mother was thoroughly scientific, and his mother was ultimately acquitted. A full scholarly account of the trial is given in Ulinka Rublack, *The Astronomer & the Witch: Johannes Kepler's Fight for his Mother*.

to explain both the number of planets and the orbital arrangement of the planetary system. This theme persists and is significantly expanded in the *Harmonice Mundi*. In the epigram of Book IV, Kepler quotes a passage from Proclus's commentary on the first book of Euclid's *Elements* that is helpful for understanding the Keplerian view of harmony. The passage deals with the use of mathematics to elucidate harmony in the natural world:

> It [mathematics] furnishes everything that is important for the contemplation of nature, declaring the most splendid order of the ratios, according to which the whole of this universe has been constructed, and the analogy of the proportions, which connects together everything in the world, as Timaeus says somewhere, and which restores friendship between things which are in conflict, and relations and mutual affection between those which are widely separated.[4]

The universe, according to Kepler, is arranged in orderly proportions that are accessible by mathematics, particularly geometry and harmonics. Helpfully, Aviva Rothman notes that:

> Kepler's understanding of harmony…privileged a number of components that were either absent or undervalued in most theories of harmony before the sixteenth and seventeenth centuries, as he understood them: polyphony, or the ability of multiple voices to express themselves; consonances that were true to experience rather than merely to a truth determined mathematically; variety of harmonic forms; and dissonance itself as both inevitable and central to the ultimate experience of harmony.[5]

Thus, for Kepler, the concept of harmony included yet also transcended the sensible consonances of music; he saw the mathematical proportions of geometry as pure archetypal harmonies that

---

4 Johannes Kepler, *The Harmony of the World*, 281.

5 Aviva Rothman, *The Pursuit of Harmony: Kepler on Cosmos, Confession, and Community*, 24.

underlie the cosmos. Again, such proportions are only perceived by a rational mind; he says that "what proportion is without the action of the mind is something which cannot be understood in any way."[6]

Harmony dominated Kepler's thought for the rest of his life, even when financial necessity required him to spend extended periods of time on what he considered mundane mathematical and astrological work.[7] Max Caspar elaborates:

> [A]t every opportunity thoughts from the extensive complex questions of his *Harmony* emerged in Kepler's mind. They ripened in him as his knowledge grew in extent and content. The fruits of his works in astronomical, mathematical, philosophical domains filled and nourished his supply of harmonic ideas, served them for cleansing, correction, widening and deepening, offered supports and presented new combinations of thoughts.[8]

In 1607, Kepler had received a Greek edition of Ptolemy's *Harmonics*, which he had long been eager to examine; it stimulated his own thinking about harmony, although he ultimately disagreed with Ptolemy's approach to realizing the Pythagorean music of the spheres.[9] Thus, Kepler's idea of harmony had developed over several years by the time he wrote the *Harmonice Mundi*. As previously indicated, he drew a clear distinction between harmonies of the senses (such as those heard in music) and pure harmonies. Harmony itself, he believed, is fundamentally about the soul's perception of a relation, which is an object of reason.[10] As Caspar puts it, "by its own nature the soul contains the pure harmonies as prototypes or paradigms of the harmonies of the senses."[11]

---

6   Kepler, *The Harmony of the World*, 10.

7   For Kepler's views on astrology as a discipline, see Endnote 10.

8   Max Caspar, *Kepler*, 266.

9   Bruce Stephenson, *The Music of the Heavens: Kepler's Harmonic Astronomy*, 99–100, 111.

10 Caspar, *Kepler*, 269.

11 Ibid., 270.

In the years following the publication of the *Mysterium*, Kepler had become increasingly aware of the discrepancies between his polyhedral model of planetary orbits and the best available observational data. Rather than scrapping his theory altogether, he sought a mathematical explanation that would allow him to adequately modify the geometrical model. Largely inspired by Ptolemy's *Harmonics*, which assigned musical scales to the motions of the spheres, Kepler set out to construct a mathematical harmonic theory based upon the fastest (when closest to the sun) and slowest (when furthest from the sun) angular velocities of the planets. Rather than using arithmetic to determine the consonance or dissonance between the note interval ratios, the mathematical harmonies Kepler used as the basis for the orbital eccentricities and periods of the planets were derived from polygonal geometry. Unfortunately, Kepler's scheme was incredibly convoluted, requiring extensive mathematical gymnastics in order to achieve only an approximate correspondence between the notes of just intonation and the celestial movements. However, he was reasonably satisfied with his scheme.

In the course of his harmonic investigations, which were fueled by his archetypal philosophy and convictions about cosmic harmony, Kepler made a discovery that is now recognized as an essential milestone in the rise of modern astronomy. This was his Third Law of planetary motion (discovered on May 15, 1618), which showed a constant mathematical relationship between the orbital period (P, the Earth years required for a planet to orbit the sun once) and that planet's mean distance from the sun (a, expressed in astronomical units).

$$P^2 = a^3$$

Thus:

$$P^2/a^3 = 1$$

This formula works for all of the planets, despite the fact that they have different orbital periods and different mean distances from the sun. Kepler delighted in this law because it demonstrated a mathematical harmonization of motion and distance among all planets—which is why it is also known as the harmonic law.

KEPLER'S 3rd LAW EXAMPLE

The square of a planet's orbital period, P, is directly proportional to the cube of its mean distance from the Sun, a.

Planet	P (year)	P^2	a (AU)	a^3	P^2/a^3
Mercury	0.24	0.06	0.39	0.06	0.99
Venus	0.62	0.38	0.72	0.38	1.02
Earth	1.00	1.00	1.00	1.00	1.00
Mars	1.88	3.53	1.52	3.51	1.01
Jupiter	11.86	140.66	5.20	140.61	1.00
Saturn	29.46	867.89	9.58	879.22	0.99

Figure 6.1 *Kepler's 3rd Law Example*

As astronomer and historian of science Owen Gingerich explains, Kepler regarded the Third Law as "a clear and accurate manifestation of the more fundamental principles underlying the cosmos—both physical and archetypal."[12] In Kepler's own words, taken from the introduction to Book V of the *Harmonice*:

> I have brought that discovery into the light, and have most truly grasped beyond what I could have ever hoped: that the whole

---

12   Owen Gingerich, *The Eye of Heaven: Ptolemy, Copernicus, Kepler*, 354.

nature of harmony, in its full extent, with all its parts...is to be discovered among the celestial motions. It is to be discovered indeed not in the way which I had mentally conceived (and this is not the least part of my joy), but in a totally different way, and also at the same time a quite outstanding and perfect way.[13]

By "way which I had mentally conceived," Kepler is referring to his earlier theory about the interpolation of the Platonic solids in between the planetary orbits and explaining that he has now become convinced that the archetypal plan in its fullness has much more to do with the harmonic relations of the orbits. Included in these relations is the Third Law, a fact that is clear from his words later in Book V:

Again, therefore, a part of my *Secret of the Universe* [*Mysterium Cosmographicum*], put in suspense 22 years ago because it was not yet clear, is to be completed here, and brought in at this point. For when the true distances between the spheres were found, through the observations of Brahe, by continuous toil for a very long time, at last, at last, the genuine proportion of the periodic times to the proportion of the spheres—

Only at long last did she look back at him as he lay motionless,
But she looked back and after a long time she came;[14]

And if you want the exact moment in time, it was conceived mentally on the 8th of March in this year one thousand six hundred and eighteen, but submitted to calculation in an unlucky way, and therefore rejected as false, and finally returning on the 15th of May and adopting a new line of attack, stormed the darkness of my mind.[15]

---

13 Kepler, *The Harmony of the World*, 389–390.

14 Here, according to the translators' annotation, Kepler quotes from Virgil, *Eclogue* I, 27 and 29.

15 Kepler, *The Harmony of the World*, 411. For a detailed resolution of a discrepancy that exists in the scholarly literature, see Endnote 11.

In the introduction to Book V, which Kepler rewrote shortly after his discovery of the Third Law, he describes his intellectual and spiritual ecstasy:

> Now…a very few days after the pure Sun of that most wonderful study began to shine, nothing restrains me; it is my pleasure to yield to the inspired frenzy, it is my pleasure to taunt mortal men with the candid acknowledgment that I am stealing the golden vessels of the Egyptians to build a tabernacle to my God from them, far, far away from the boundaries of Egypt.[16]

Like Augustine, Kepler alludes to the biblical story of the Exodus, in which the Israelites took silver and gold from the Egyptians, just before their escape from bondage, and later used these treasures to build the Tabernacle (Exodus 25:1–8). Presumably, the "Egyptians" in Kepler's case are the non-Christian Pythagorean-Platonic philosophers from whom he gleaned valuable intellectual riches that could be pressed into service to God's kingdom.[17]

It should be noted that although Kepler's Book V introduction does not explicitly refer to the harmonic law (the Third Law), there is good scholarly consensus that this is precisely what inspired his rapturous enthusiasm.[18] Caspar waxes poetic about Kepler's attitude towards the discovery:

> The magic of the word harmony transported him to another, a pure, paradisiacal world. Just as the Greeks, who created the word, set the idea of harmony in the center of their cosmology and from there sought to advance to the root of existence, so was all his thinking filled and ruled by this idea. The wall of sense matters became transparent like glass. Here in the idea of harmony he found unity in the many-sidedness of phenomena, the

---

16 Kepler, *The Harmony of the World*, 391.

17 For a discussion on the early history of the Egyptian treasures metaphor, see Endnote 12.

18 See discussion in Bruce Stephenson, *The Music of the Heavens*, 129.

essential kernel of nature, the supporting principle of regulation, which really and actually and in fact makes the world into the loveliest possible, the key to the understanding of relationship, connecting the created spirit with the original spirit, with God.[19]

Kepler saw the Third Law as the ultimate affirmation of his philosophy of nature. This is an essential point, because it highlights the essence of Kepler's idea of the tripartite harmony that relates the mind of God ("original spirit") to the mind of man ("created spirit") by way of the created order. At this point, he had already learned that his geometrical model was not entirely consistent with the observational data; had he lived to see his polyhedral theory and his musical ratios fall permanently into the dustbin of science history, the affirmation of harmony that he saw in his mathematical Third Law would have remained, even (or even especially) with its later Newtonian modifications.[20] What Kepler ultimately valued, it seems more than fair to say, was the underlying mathematical ordering of the world—with all its interconnections between structure and motion—that mankind is able to discern. As will be seen in later chapters, cosmic rationality and harmony goes far above and beyond what Kepler could have ever imagined.

There can be no doubt that the Pythagorean-Platonic tradition was a significant inspiration for Kepler's work in the *Harmonice*. For instance, a major point of Keplerian philosophy is that harmony can only exist when there is a rational soul to perceive it. In other words, there must be an intellect with the inherent power to see that things fit together in a pleasing way. Bruce Stephenson writes:

---

19 Caspar, *Kepler*, 267.

20 Kepler, *Mysterium Cosmographicum*, 71. Kepler's own discussion of his growing understanding of the inexactitude of the polyhedral model can be found in his footnotes to the Preface of the revised edition of the *Mysterium Cosmographicum*, published twenty-five years after the first edition (1621).

> [*Harmonice Mundi*] treated harmony in its mathematical, musical, astrological, and astronomical aspects, culminating in an analysis of the harmonies in the motions of the planets. Treating all these diverse fields of knowledge under the single concept of "harmony" was not only permissible but necessary, in Kepler's opinion, for all these forms of harmony had the same mathematical basis. That basis was to be found in geometrical relations between physical quantities rather than in the purely numerical relations between integers on which the Pythagorean harmonic theory was based. For a relation to be harmonic, Kepler thought, its beauty had to be perceived by a soul, which might be human or otherwise.[21]

For Kepler, the human soul, made in the image of God, was the third part of the tripartite harmony of archetype (rational plan of creation in the creating Mind), copy (the material manifestation of the archetype), and image (the soul that apprehends harmonious relationships). The mathematical plan for creation is made manifest in the material creation, and the human intellect is designed to recognize its inherent harmonies and thereby gain glimpses into the very mind of God.

Kepler's philosophy of nature is quite Pythagorean-Platonic, as can be seen by way of comparison with the *Timaeus*. Timaeus explains that the cosmos was created according to "a pattern intelligible and always the same" and that this copy "was only the imitation of the pattern, generated and visible."[22] He also says that the human soul was created with the sensory and intellectual faculties necessary for the appreciation of harmonies: "God invented and gave us sight to the end that we might behold the courses of intelligence in the heaven, and apply them to the courses of our own intelligence which are akin to them."[23] By "courses of intelli-

---

21 Stephenson, *The Music of the Heavens*, 4–5.

22 Plato, *Timaeus*, 456.

23 Ibid., 455.

gence," Plato means the harmonious motions of the heavens and by "courses of our own intelligence" he means the corresponding harmonies in the human soul. The conceptual similarity with Kepler's sentiments is striking. In Chapter I of Book III of the *Harmonice*, Kepler specifically refers to the *Timaeus*:

> ...the operations and motions of [the heavenly] bodies, which imitate the harmonic proportions, are on the side of the soul and the mind, assigning them a cause for their delight in consonance...Indeed the philosophy of Timaeus the Locrian on the composition of the soul from harmonic proportions...was refuted by Aristotle in the sense conveyed by the actual words; but I should not dare to affirm that there is nothing lurking in those writings but what the actual words convey. On the contrary I think no-one will deny that the author at least holds what I here ascribe to him, that it is Mind or the human intellect by the judgement or instinct of which the sense of hearing discriminates pleasant, that is consonant proportions from the unpleasant and dissonant, especially if he ponders carefully that proportions are entities of Reason, perceptible by reason alone, not by sense, and that to distinguish proportions, as form, from that which is proportioned, as matter, is the work of Mind.[24]

What Kepler seems to suggest here is that the passage from the *Timaeus* about the soul is not to be interpreted at literal face value, but rather metaphorically. He takes Plato to mean that the soul is attuned to the harmonies in the world. Later, in Chapter II of Book IV, Kepler explains that sensitivity to harmonic proportions is intrinsic to the human soul, which is divided into a lower, sensible part that simply recognizes harmony, and a higher, contemplative part that seeks and grasps "the actual proportions in abstract quantities" that are fundamental to them. Mathematical harmonies "elicit intellectual things which are previously present within, so

---

24 Kepler, *The Harmony of the World*, 150.

that things now in actuality shine forth in the soul which were hidden in it before, as if under a veil of potentiality."[25] Caspar points out that, on this point, Kepler "is a follower of Plato, according to whose theory the human mind learns all mathematical ideas and figures, all axioms, all solutions about these things out of itself; by the physical signs it only remembers that which it knows out of itself...the mathematical ideas are the essence of the soul and inversely the soul is the essence of them."[26] For Kepler, this imprinted nature of the soul is the direct result of man being created in the image of God, having a "natural light" in tune with the mind of our Maker. According to Rothman, "The Platonic doctrine of recollection clearly signaled, to Kepler, that the Christian God had somehow imbued knowledge of fundamental truths—for Kepler, geometric objects—in the minds of his creatures...Moreover, with Kepler's mention of natural light as the way in which human knowledge was rooted in the divine, Kepler referenced not only Plato and Proclus but also Augustine and his followers."[27]

In Book II of the *Harmonice*, Kepler proceeds through a series of geometrical axioms and propositions and then from a discussion on polygons and the various possible tessellations to the construction of polyhedra. For the latter, he explains how polygons (plane figures) come together to form the multiple faces of each three-dimensional polyhedron. About the Platonic solids he writes, "The most perfect regular congruences of plane figures to form a solid figure are five in number...These are the five bodies which the Pythagoreans and Plato, and Euclid, were accustomed to call the world figures."[28] This is akin to Plato's approach to the construction of the solids in sections 54 and 55 of the *Timaeus*. Kepler

---

25   Kepler, *The Harmony of the World*, 307.

26   Caspar, *Kepler*, 270. Here, Caspar refers to the doctrine of recollection, which Plato explores in the *Meno*.

27   Rothman, *The Pursuit of Harmony*, 49.

28   Kepler, *The Harmony of the World*, 113–114.

goes on to explain the polyhedral theory of elements (earth, fire, water, air), which he says he is taking from Aristotle. There is no indication that Kepler believes this theory, and in fact, he completely ignores it in *De Nive Sexangula* (*The Six-Cornered Snowflake*), a work that is recognized as a precursor for modern crystallography. That would have been the perfect place to discuss the polyhedral element theory, since he uses that short book to explore the formal effects of geometry at the fundamental level of nature. Yet, Kepler regards the polyhedral theory of elements as an acceptable analogy, because it is resonant with his archetype/copy philosophy of nature which (he is still convinced) involves the Platonic solids.[29]

Kepler goes on to argue that Aristotle's cosmology cannot include archetypes and therefore is incompatible with the Christian paradigm. He says that the archetypes do not work with the universe of Aristotle, "who did not believe that the World had been created and thus could not recognize the power of these quantitative figures as archetypes, because without an architect there is no such power in them to make anything corporeal."[30] The archetypes are merely models; they do not have the power in themselves to cause any effects in the material realm; an architect is necessary to make them manifest. Kepler then explains that the archetypes are fully compatible with Christian theology, "since our Faith holds that the World, which had no previous existence, was created by God in weight, measure, and number, that is in accordance with ideas coeternal with Him."[31] It is worth reiterating that the phrase "weight, measure, and number" echoes many pagan, Jewish, and Christian writers who drew from the Pythagorean-Platonic tradition in which Kepler's thought was steeped.[32]

---

29 Ibid., 115.

30 Ibid.

31 Ibid.

32 As seen in Chapters 2 and 3, examples include Nicomachus of Gerasa, Philo of Alexandria, the *Book of Wisdom*, and Augustine of Hippo.

There are other interesting ways in which Kepler's natural philosophy reflected the influence of the Pythagorean-Platonic tradition. For example, in Book V of the *Harmonice*, Kepler makes an explicit connection between his understanding of creation and Platonism. Alongside an extended quotation from Proclus's *Commentary on the First Book of Euclid's Elements*, he includes a marginal notation about the *Timaeus* which reads, "In the Timaeus, which is beyond all hazard of doubt a kind of commentary on the first chapter of Genesis, or the first book of Moses, converting it to the Pythagorean philosophy, as is readily apparent to the attentive reader who compares the actual words of Moses in detail."[33] Although Kepler's idea about Plato's *Timaeus* being intended as a commentary on the Genesis creation account is radical (and, in modern Plato scholarship, dubious), it nevertheless seems significant that Kepler classifies the *Timaeus* in this manner, in light of the fact that he himself adopted some of the same concepts in his archetypal account of the creation.

In her highly regarded evaluation of Kepler's geometrical approach to cosmology, J. V. Field writes that "Kepler's use of mathematics in his cosmological works is very like Plato's use of mathematics in *Timaeus*," yet unlike Plato, "Kepler constructs not a vague metaphysical description of the cosmos but a testable mathematical model."[34] She notes that some of the Platonic influence on Kepler's thought came from Proclus's commentaries, which Kepler quotes repeatedly in the *Harmonice*. Proclus refers to a passage in section 32 of the *Timaeus* which reads, "And for these reasons, and out of such elements which are in number four, the body of the world was created, and it was harmonised by proportion…"[35]

---

33 Kepler, *The Harmony of the World*, 301.

34 J. V. Field, *Kepler's Geometrical Cosmology* (Chicago: University of Chicago Press, 1988), 16.

35 Ibid., 99.

He then alludes to the mathematical descriptions used by Timaeus in sections 53 through 61:

> [Timaeus] everywhere describes his reflections on the nature of the whole universe in mathematical terms, depicts the origin of the elements in numbers and figures, and states that their powers and properties, and their effects, were taken from these, the acute and obtuse among the angles, and the rough and smooth among the sides, and so forth, establishing the causes of all kinds of mutations.[36]

Proclus, following Plato, emphasizes both the mathematical and aesthetic value of symmetry and simplicity.[37] These principles were key in Kepler's cosmological convictions, including his early acceptance of Copernicanism as well as his lifelong search for an underlying mathematical cosmic harmony.

In the years surrounding the publication of the *Harmonice*, Kepler was also in the process of publishing the individual books of his *Epitome Astronomiae Copernicanae*. This defining work, which truly represents the "new astronomy," reads much like a catechism, with questions followed by detailed answers meant to show the superiority of the heliocentric system. Working from a foundation composed of his three planetary laws, Kepler meticulously constructed a model that supplanted the epicycle-ridden Aristotelian-Ptolemaic system. The title is a misnomer; as Caspar explains, "Kepler had left Copernicus, also, far behind. He had created a new science; he had undertaken to conceive and to explain the motions physically; he had offered the first textbook of celestial mechanics."[38] The polyhedra and harmonic principles are not en-

---

36 Kepler, *The Harmony of the World*, 281. Kepler was also familiar with Proclus's commentary on the *Timaeus*, which he quotes in Book III of the *Harmonice* during a discussion on the Pythagorean tetractys. See *The Harmony of the World*, 133.

37 J. V. Field, *Kepler's Geometrical Cosmology*, 99.

38 Caspar, *Kepler*, 294.

tirely absent from the *Epitome*, but they do not have primacy of place. Kepler is still concerned with archetypal (formal) causes, but the *Epitome* emphasizes the natural (efficient) causes at work in celestial physics. It should be noted that Kepler did not entirely dispense with a type of "world soul" to account for celestial motions. For one thing, he still assumed that such an animating principle was necessary for the ongoing axial rotation of the earth and sun. Perhaps it is reasonable to conclude that Kepler uses the world soul idea as a placeholder, one awaiting supplantation by a natural, mechanical explanation. This would be consistent with his previous tendencies. Ultimately, however, Kepler's treatise is truly the monumental transition from ancient cosmology to a bona fide mechanical astronomy; rather than merely saving the appearances by providing a geometrical model that accounted for the observational data, Kepler sought to also provide corresponding *physical explanations* for celestial dynamics. To the question of why the *Epitome* typically does not receive its due credit, Caspar replies that

> the *Epitome* ranks next to Ptolemy's *Almagest* and Copernicus' *Revolutiones* as the first systematic complete presentation of astronomy to introduce the idea of modern celestial mechanics founded by Kepler. Not the least reason that the conventional presentations of the history of science do not concede this place to the work is the fact that it is too little known…But the title…also bears the blame…[it] gives no inkling that Kepler had erected an entirely new structure on the foundation of the Copernican theory, that he had rescued the Copernican conception, at that time disputed and little believed, and helped it to break through by introducing his planet laws and by treating the phenomena of the motions physically."[39]

Moreover, Kepler's modesty in his work, along with the fact that it indeed fell short in certain ways that Newton's work did not,

---

39 Caspar, *Kepler*, 297.

likely contributed to the *Epitome*'s underappreciation by many historians of science.

Excursus: Galileo and Newton

Before undertaking a more detailed study of the philosophical and theological aspects of Kepler's thought, it seems fitting to first situate his discoveries within the broader scientific revolution by briefly explaining their connection to the work of his most renowned contemporary, Galileo Galilei (1564–1642), and his main successor, Sir Isaac Newton (1643–1727). All three men played major, interconnected roles in the emergence of early modern science; like Kepler, Galileo and Newton drew positive theistic implications from their increasing knowledge of the natural world, and some of these implications bore the distinctive gloss of the Pythagorean-Platonic tradition.

In 1609, in order to aid his astronomical observations, Galileo Galilei began building a series of double-lens telescopes, each one improving upon the previous in its level of magnification. While he was not the original inventor of the device (at the time called a "perspiculum"), his ingenuity resulted in a far more sophisticated and powerful instrument than the simple spyglass that had inspired him. In his work entitled *The Sidereal Messenger*, Galileo describes his telescopic discoveries, including the existence of the four satellites of Jupiter that essentially disproved the Ptolemaic assumption that there are no celestial bodies that orbit around any planet other than the earth.[40] He also details his observations of the phases of Venus, which he understood as evidence that Venus orbits the sun and not the earth.[41] (The Venusian phases constituted empirical

---

40 Ibid., 191.

41 Thomas Kuhn, *The Copernican Revolution*, 223–224.

support for the heliocentric model, but not definitive proof. At the time, the Tychonic geo-heliocentric system in which the planets revolve around the sun and the sun revolves around the earth provided as good of an explanation for the Venusian phases as the one offered by the Copernican model.)[42] As a result of his findings, Galileo became an ardent advocate for heliocentrism and argued his case in *Dialogue on the Two Chief World Systems.*

Unfortunately, and for whatever reason, Galileo did not incorporate Kepler's planetary laws into his work even though he was well aware of them; thus, the Galilean model did not represent the best astronomy of the time.[43] It was still mired in what Galileo thought of as the "natural" circular motions of the Aristotelian-Ptolemaic system, even though Kepler's First Law (elliptical orbits) was a superior explanation of the observational data. As Caspar argues:

> In historical accounts it is repeatedly stated that it was Galileo who founded the Copernican theory physically. While fully appreciating Galileo's accomplishments in the domain of mechanics, it must still be emphatically pointed out that he completely failed to comprehend the idea of a celestial mechanics. In none of his works did he take notice of Kepler's planet laws although he certainly knew them. Not once in his famous Dialogue about the systems of the world, which appeared a quarter of a century later, did he speak of them, although they surely should have played a central part...So it was Kepler first of all, not Galileo, who freed astronomy from the bonds of Aristotelian physics.[44]

---

42 Thomas Kuhn, *The Copernican Revolution*, 224.

43 Multiple sources affirm this. For example, a letter dated July 21, 1612 from Frederico Cesi to Galileo discusses Kepler's elliptical orbit theory as a matter of common knowledge. See Caspar, *Kepler*, 137.

44 Caspar, *Kepler*, 136.

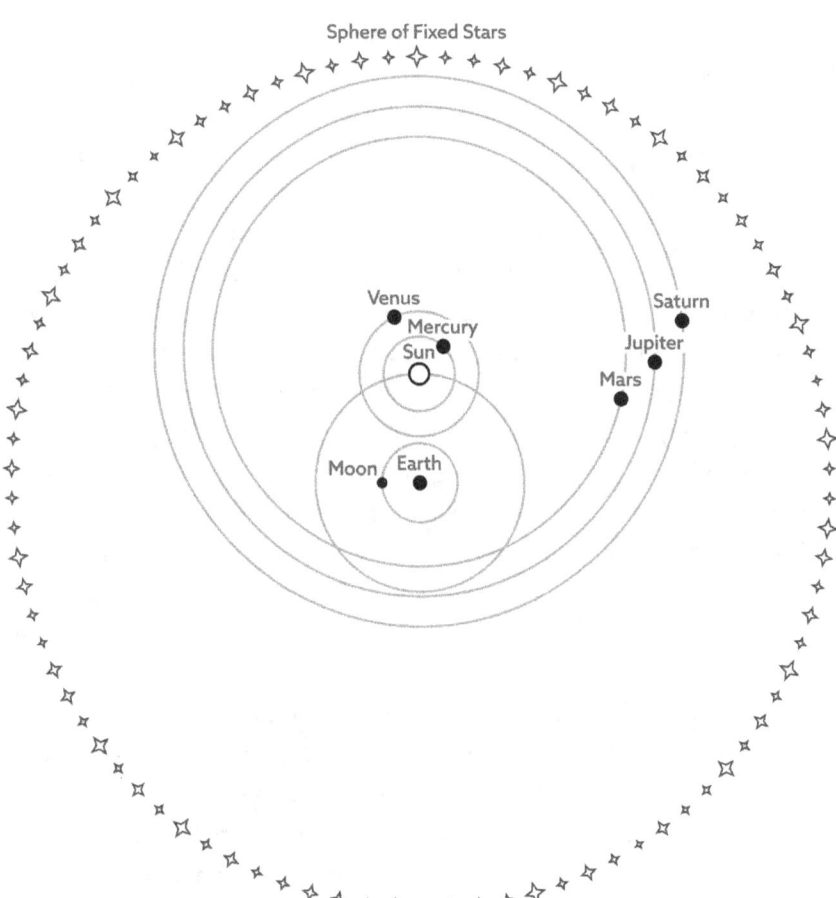

Figure 6.2 *The Tychonic System*

Indeed, Galileo presented an inferior physical cosmology and failed to achieve what Kepler accomplished. Yet, the fact remains that his integrative approach to astronomy, involving both telescopic observations and a mathematical framework for gaining knowledge about the world, was enormously significant for subsequent scientific inquiry. To give credit where credit is due, it was Kepler who published the theory of telescopic optics (*Dioptrice*, 1611) and proposed a superior arrangement of lenses, and the Keplerian telescope was adopted as the instrument of choice in the practice of astronomy.[45]

Galileo's application of mathematics to the physical world was not limited to astronomy; he also experimented with falling objects and projectile motion. In his 1638 *Discourses on Two New Sciences*, he documents his experimental simulation of a free-fall using balls and inclined planes (to slow down the motion enough to be timed). His contribution to the collective knowledge about how gravity affects freely falling bodies demonstrated that the distance an object has fallen is proportional to the square of the time it took to travel said distance.[46] Also, in his work on projectile motion, Galileo described the trajectory of a launched cannonball as a combination of the straight, forward motion out of the cannon and the downward pull of gravity, which together take the ball along a parabolic curve.[47] This work in terrestrial physics, combined with Kepler's planetary laws, was essential to Newtonian physics. It should not pass without notice, however, that Kepler's *Epitome* had already discussed problems of terrestrial physics, such as why (if the earth is rotating) a vertically thrown stone comes down to land at the same place and a launched cannon ball has the

---

45 Owen Gingerich, "Creative Revolutionaries: How Galileo and Kepler Changed the Face of Science," *Euresis* 2 (Winter, 2012): 10.

46 James Hannam, *The Genesis of Science: How the Christian Middle Ages Launched the Scientific Revolution* (Washington, DC: Regnery Publishing, 2011), 339.

47 Ibid., 342.

same projectile range whether it is fired to the east or to the west. Kepler had already arrived at the answers, but Galileo treated them in more mathematical detail and is customarily credited.[48]

Like Kepler, Galileo viewed his natural philosophy as being supportive of Christian theism. In a 1613 letter to former student Benedetto Castelli, a Benedictine monk who served as professor of mathematics at Pisa, Galileo discusses the "two books" philosophy of divine revelation: "For the Holy Scripture and nature both equally derive from the divine Word, the former as the dictation of the Holy Spirit, the latter as the most obedient executrix of God's commands."[49] Later, in a 1615 letter to the Grand Duchess Christina—a letter that echoes the one to Castelli to a great extent—Galileo says, "For the Holy Scripture and nature derive equally from the Godhead, the former as the dictation of the Holy Spirit and the latter as the most obedient executrix of God's orders…God reveals Himself to us no less excellently in the effects of nature than in the sacred words of Scripture."[50] Then, in his 1623 work, *The Assayer*, he alludes to the book of nature metaphor in what has become a widely known statement:[51]

> Philosophy is written in this all-encompassing book that is constantly open before our eyes, that is the universe; but it cannot be understood unless one first learns to understand the language and knows the characters in which it is written. It is written in mathematical language, and its characters are triangles, circles, and oth-

---

48   See discussion in Caspar, *Kepler*, 294–295.

49   Galileo Galilei, *Letter to Castelli (1613)*, in *The Essential Galileo*, ed. and trans. Maurice A. Finocchiaro (Indianapolis: Hackett Publishing, 2008), 9, 104.

50   Galileo Galilei, *Letter to the Grand Duchess Christina*, in *The Essential Galileo*, 116. It has been suggested that the Duchess never actually read this letter, that it was a rhetorical exercise.

51   The 1959 Disney feature film entitled *Donald in Mathmagic Land* explores the mathematical rationality of nature, and the ending frame shows and narrates a portion of this quote. The film was also my introduction to the Pythagoreans. Hence, I can trace my inspiration for this project back to the 1980s Disney Channel.

er geometrical figures; without these it is humanly impossible to understand a word of it, and one wanders around pointlessly in a dark labyrinth.[52]

According to Galileo, creation is open to the observation and analysis of man, who can comprehend the inherently mathematical universe if he first learns the mathematical arts.

For Galileo, man's aptitude for mathematical reasoning suggests a relationship to the divine source of the universe. He directly expresses his sympathy with this aspect of Pythagorean-Platonic philosophy in his *Dialogue*, where he says: "That the Pythagoreans held the science of numbers in high esteem, and that Plato himself admired human understanding and thought that it partook of divinity, in that it understood the nature of numbers, I know very well, nor should I be far from being of the same opinion."[53] This is a noteworthy statement in that it resonates powerfully with Kepler's understanding of the tripartite harmony among the mind of God, nature, and the mind of man.

In his magnum opus, *Mathematical Principles of Natural Philosophy* (*Principia* for short), a work still hailed as one of the most important scientific treatises of all time, English mathematician and astronomer Sir Isaac Newton (1643–1727) presented his theory of universal gravitation. The idea behind his theory was that the laws governing terrestrial objects are the same laws involved with the dynamics of the planetary orbits. The genius of Newton's work was that it established a set of universal principles that could be used to explain a broad range of phenomena, whether astronomical or terrestrial. In Newtonian physics, the same laws at work with an apple falling from a tree and the trajectory of a launched cannonball could be applied to the revolutions of the planets around the

---

52 Galileo Galilei, *The Assayer*, in *The Essential Galileo*, 183.

53 Galileo Galilei, *Dialogue Concerning the Two Chief World Systems*, trans. Stillman Drake (Berkeley: University of California, 1962), 11.

sun and the moon around the earth. Newton's cosmology was the first to incorporate René Descartes's conception of inertia, which Kepler's model had unfortunately lacked.[54] However, as Kuhn succinctly explains, the Newtonian system demonstrated the veracity of Kepler's three planetary laws.[55] Newton had

> succeeded in working out mathematically the rate at which a planet must "fall" toward the sun, or the moon toward the earth, in order to remain in a stable circular orbit. Then, having discovered how this mathematical rate of fall varied with the planet's speed and with the radius of its circular orbit, Newton was able to deduce two immensely important physical consequences. If the speeds of the planets and their orbital radii were related to each other by Kepler's Third Law, then the attraction that drew planets to the sun must…decrease inversely as the square of the distance separating them from the sun. A planet twice as far from the sun would require only one-fourth the attractive force to remain in its circular orbit at its observed speed…The same inverse square law that governed the attraction between sun and planets would, he found, account quite well for the difference in the rate at which the distant moon and a nearby stone fell to the earth. Thirteen years later…he generalized his results still further and showed that an inverse-square law would account precisely for both the elliptical orbits specified by Kepler's First law and the speed variation described in the Second.[56]

One cannot help but imagine how much Newton's law of gravitation, which was the core of a unifying and thoroughly mathemat-

---

54 Kuhn, *The Copernican Revolution*, 247.

55 Eventually, with improved telescopic quantitation, it was demonstrated that planetary motion deviates a little from Kepler's laws due to the minor gravitational pull the planets exert upon one another. Newtonian physics was able to explain and predict these deviations. See discussion in Kuhn, *The Copernican Revolution*, 261ff.

56 Kuhn, *The Copernican Revolution*, 256.

ical physics, would have delighted Kepler, who had been inching towards such discoveries only a few decades before.

Newton's work sounded the death knell for the Aristotelian universe; the new system provided unprecedented comprehensiveness and coherence. The so-called Copernican Revolution was complete, and the laborious transition provided a range of new mathematical tools for natural philosophers. Kuhn writes, "The same Newtonian principles which, by providing an economical derivation and a plausible explanation of Kepler's Laws, closed the astronomical revolution also supplied astronomy itself with a host of powerful new research techniques."[57] Unfortunately and for reasons that can only be speculative, Newton failed to credit Kepler; as physicist and historian of science Gerald Holton explains, "Newton manages to remain strangely silent about Kepler throughout Books I and II of the *Principia*, by introducing the Third Law anonymously as 'the phenomenon of the 3/2th power' and the First and Second Laws as 'the Copernican hypothesis.'"[58]

Newton, like Kepler and Galileo, drew attention to the powerfully positive theistic implications of his work. In the second edition of his *Optiks* (1717), he says that the main goal of natural philosophy is to formulate convincing arguments for the existence of God. He puts this belief into practice in the "General Scholium" of the *Principia*, where he writes, "This most beautiful System of the Sun, Planets, and Comets, could only proceed from the counsel and dominion of an intelligent and powerful being."[59] For him, this being was none other than the God of Christianity

---

57 Kuhn, *The Copernican Revolution*, 261.

58 Gerald Holton, *Thematic Origins of Scientific Thought* (Cambridge: Harvard University Press, 1988), 60. Owen Gingerich suggests that Newton likely never read Kepler's *Astronomia Nova* or *Harmonice Mundi*, but that Newton did own a copy of Mercator's *Institutionum Astronomicarum* (1676), which contained the Third Law. See Gingerich, "Kepler and the Laws of Nature," 21.

59 Isaac Newton, General Scholium to the *Principia Mathematica*, 3rd ed., 1726.

(though some of Newton's theology was decidedly unorthodox, such as his denial of the doctrine of the Trinity). Newton saw his work as child's play compared to the remaining mysteries of the Creator's work. On one occasion he is said to have remarked, "I do not know what I may appear to the world, but to myself I seem to have been only like a boy playing on the sea-shore, and diverting myself in now and then finding a smoother pebble or a prettier shell than ordinary, whilst the great ocean of truth lay all undiscovered before me."[60]

Like Kepler and Galileo before him, Newton recognized the fundamentally mathematical structure of nature and saw a connection between this and the intellect of mankind. In some of his unpublished scientific papers, it is evident that he regarded the human intellect as resembling God's, though only to a very limited degree; he writes that "the analogy between the Divine faculties and our own is greater than has formerly been perceived by Philosophers. That we were created in God's image holy writ testifies."[61] Newton seems to have considered his own work to be an investigation into how much of the divine mind could be discovered through the practice of natural philosophy.[62] He understood that natural revelation, in order to function as revelation, requires creatures equipped to comprehend it; human beings perceive the divine rationality revealed in nature because they are made in the image of God. Another facet of his argument for this kinship seems to be based upon the fact that God created the world by an act of divine will, and similarly, man is endowed with faculties (of similar kind

---

60 Joseph Spence and John Underhill, *Spence's Anecdotes, Observations, and Characters of Books and Men* (London: W. Scott, 1890), 70. See Endnote 13 for an example of this statement's poetic legacy.

61 A. R. Hall and M. B. Hall, ed. *Unpublished Scientific Papers of Isaac Newton* (Cambridge: Cambridge University Press, 1962), 141–142.

62 Snezana Lawrence and Mark McCartney, ed., *Mathematicians and their Gods* (Oxford: Oxford University Press, 2015), 124.

but lesser degree) that enable him to act and create by a movement of personal free will, though he cannot produce anything out of nothing the way the Almighty did.[63]

---

63 A. R. Hall and M. B. Hall, *Unpublished Scientific Papers of Isaac Newton*, 141.

CHAPTER 7

The Scientific Fruitfulness of Kepler's Harmonies

In several respects, Kepler was truly a transitional figure. He emphasized the importance of unity in explanations for natural phenomena, whether celestial or terrestrial, as well as the primacy of empirical data over philosophical precommitments (such as the circular orbits of the Aristotelian-Ptolemaic model that even Galileo clung to). His life's work was driven by the theological belief that the rational plan of creation, pre-existent in the mind of God, is stamped upon the world in a way that makes it accessible to the mind of man, a creature made in God's image. In terms of the harmonization of the book of nature with the book of Scripture, Kepler was an accommodationist; he understood passages related to natural phenomena as reflections of everyday sensory perception rather than literal facts of natural philosophy. Thus, Kepler's thought remains highly relevant to contemporary natural theology.

As aforementioned, Kepler's work was pivotal in the shift from ancient to modern astronomy. His substitution of physical forces for the purely geometrical cosmologies of ancient astronomers was revolutionary—he produced a true *physica coelestis*.[1] When he postulated a magnetic force emanating from the sun to explain the orbital motion of the planets, he was, as Owen Gingerich ar-

---

1 Pietro Daniel Omodeo, "The 'Impiety' of Kepler's shift from mathematical astronomy to celestial physics," *Annalen der Physik* 527, no. 7–8 (2015): A71.

gues, "the first scientist to demand physical explanations for celestial phenomena," and although Kepler's mathematical description of how this force diminishes with distance turned out to be only an approximation, "the important physical-mathematical step had been taken."[2] Gerald Holton concurs; he writes that "Kepler's genius lies in his early search for a physics of the solar system. He is the first to look for *a universal physical law based on terrestrial mechanics* to comprehend the whole universe in its quantitative details."[3] In the classical understanding of the world, the celestial and terrestrial were seen as two separate domains, but Kepler sought unification by hypothesizing one mechanical force that operated in both.[4] In a letter to Herwart von Hohenburg dated February 10, 1605 (while working on the *Astronomia Nova*) Kepler writes,

> I am much occupied with the investigation of the physical causes. My aim in this is to show that the celestial machine is to be likened not to a divine organism but rather to a clockwork...insofar as nearly all the manifold movements are carried out by means of a single, quite simple magnetic force, as in the case of a clockwork all motions [are caused] by a simple weight. Moreover I show how this physical conception is to be presented through calculation and geometry.[5]

Max Caspar comments, "It is Kepler's greatest service that he substituted a dynamic system for the formal schemes of the earlier astronomers, the law of nature for mathematical rule, and causal explanation for the mathematical description of motion. Thereby

---

2 Owen Gingerich, *The Eye of Heaven: Ptolemy, Copernicus, Kepler*, 309.

3 Gerald Holton, *Thematic Origins of Scientific Thought: Kepler to Einstein*, 55.

4 Ibid., 54.

5 Letter quoted in Owen Gingerich, *The Eye of Heaven*, 56. For further discussion on Kepler's search for natural causes, see Endnote 14, and for more on the clockwork metaphor, see Endnote 15.

he truly became the founder of celestial mechanics."[6] Although he never quite achieved his ultimate goal, says Holton, Kepler "succeeded at least in throwing a bridge from the old view of the world as an unchangeable cosmos to the new view of the world as the playground of dynamic and mathematical laws. And in the process he turned up, as if it were by accident, those clues which Newton needed for the eventual establishment of the new view."[7]

Unfortunately, Kepler's polyhedral/harmonic hybrid approach to elucidating the formal (archetypal) cause of the universe has led some to view him as a strange mystic whose excellent skills in mathematics and astronomy nevertheless resulted in useful discoveries in support of heliocentrism. As Bruce Stephenson explains:

> In the three and a half centuries since Kepler published his *Harmonice mundi*, it has acquired a certain notoriety as the expression of an unscientific or mystical side of his personality. To be sure, Kepler's harmonic theories were fanciful, even for their time. In part this was simply due to his boldness as a thinker; he was not afraid to draw far-reaching conclusions where others, even if they agreed in principle, might lack the intellectual courage to follow him…Kepler's theories about the intimate details of Creation seem unscientific and irrational because they presuppose a created world in which humanity occupies a uniquely important place. This is, I think, the most important reason why his harmonic theories are condemned or praised (according to the taste of the critic) as "mystical."[8]

In other words, criticism of Kepler is often motivated by metaphysical disagreement. Yet, even though Stephenson himself (like many today) seems to reject a teleological understanding of the

---

6   Max Caspar, *Kepler*, 136.

7   Gingerich, *The Eye of Heaven*, 54–55.

8   Bruce Stephenson, *The Music of the Heavens: Kepler's Harmonic Astronomy*, 8–9.

natural world, he defends Kepler's approach as wholly rational for the seventeenth century:

> If the word mystical is to have any use in historical writing, it must be applied to ideas that are less rational, more transcendent, than the commonplace ideas of their time. The *Harmonice mundi* presented no ideas opposed to reason, nor any that claimed to transcend reason. The harmonies that Kepler discerned in the heavens were entirely rational. They had been created by divine reason; they were intended, Kepler assumed, for the enjoyment of a rational soul; and they were accessible to a sufficiently diligent exercise of human reason.[9]

Kepler's expectation of conceptual unity and mathematical harmony inspired his beloved polyhedral/harmonic theory, which was eventually shown to be incorrect when the additional planets of the solar system were discovered (long after his death).[10] As a result, this part of his work is often viewed as nothing more than a fantastical historical curiosity—the last hurrah of Pythagorean mysticism in astronomy. However, it must be recognized that his Pythagorean-Platonic approach worked to his advantage in terms of his lasting contributions to science. In fact, it has been suggested by several historians of science that Kepler's path of discovery, for all its Pythagorean-Platonic oddity and various dead ends, may have been the *only* path that would have led to the unprecedented laws of planetary motion when it did. As Stephenson argues,

> It can be said of Kepler, as of very few great scientists, that what he accomplished would never have been done had he himself not done it. It is in the nature of science that one seldom can say this about important achievements. If Isaac Newton had died in the plague year the development of mechanics would have been slower but the mathematicians of the eighteenth century would

---

9 Stephenson, *The Music of the Heavens*, 9.

10 Uranus was discovered in 1781, Neptune in 1846.

eventually have obtained all his important results...With Kepler it is different. The discovery from examination of naked-eye observational reports that planets move on ellipses, and according to the area law, is so exceedingly improbable—and Kepler's manner of arriving at it was so decidedly personal—that it lies well outside the course of any inevitable development. The eventual derivation of such motion on the basis of classical physics was certain; its empirical discovery without the guidance of classical physics was really quite extraordinary.[11]

Thus, armed with only observational data and certain philosophical convictions about the rationality of the world, Kepler made his brilliant, untimely discoveries. Essentially, the two main guides for Kepler's ingenious accomplishments were a sound instinct for physics and a commitment to neo-Platonic metaphysics.[12] Biographer Kitty Ferguson elaborates:

> The comment has sometimes been made that the harmonic law was an accidental discovery in the midst of a labyrinth of worthless musical/mathematical speculation...[but] Without the underpinning of modern mathematics and the modern scientific method, the convoluted musical path Kepler took may have been the only way he could have got there. After all, he was the one who did get there. Kepler had one of the truest ears in history for the harmony of mathematics and geometry.[13]

Even though his nested polyhedra and harmonic planetary motions ultimately fell by the wayside, "They had been the odd and unlikely midwives to Kepler's 'new astronomy,'...[and] the conviction that numbers and harmony and symmetry were guides

---

11 Bruce Stephenson, *Kepler's Physical Astronomy* (Princeton, Princeton University Press, 1994), 203.

12 Gingerich, *The Eye of Heaven*, 55.

13 Kitty Ferguson, *Pythagoras: His Lives and the Legacy of a Rational Universe* (New York: Icon Books, 2008), 275.

to truth because the universe was created according to a rational, orderly plan began to be treated as a given, trustworthy enough to underpin what would later be called the scientific method."[14] Gingerich agrees: "Although the principal idea of the *Mysterium cosmographicum* [the polyhedral theory] was erroneous, never in history has a book so wrong been so seminal in directing the future of science."[15] For it was the *Mysterium* that Kepler regarded as the prequel to the *Harmonice*, and it was the work on the latter that brought him to the harmonic law. Gingerich goes on to say that "Kepler's great cosmic vision of celestial harmony—part fantasy and chimera—had indeed ultimately brought him closer to the eternal architecture of his Creator."[16]

This collective assessment of the value of Kepler's philosophy in his astronomical theorizing is consistent with the analysis of J. L. E. Dreyer (1852–1926), a Danish astronomer and historian of astronomy who penned this earlier corrective:

> Many writers have expressed their deep regret that Kepler should have spent so much time on wild speculations and filled his books with all sorts of mystic fancies. But this is founded on a misconception of Kepler's object in making his investigations of the cosmographic mystery and the harmony of the world, for even in his wildest speculations he took as his base carefully observed facts and he aimed at and obtained results of great practical value…To his determination to build up his system of polyhedra on the solid rock of thoroughly reliable and systematically-made observations was due the perseverance with which he clung to his post under Tycho…To his continued work in the same direction we owe the first and second law, and to the work on the harmony we owe the third. There is thus the most intimate connection

---

14 Kitty Ferguson, *Pythagoras*, 278.

15 Gingerich, *The Eye of Heaven*, 309.

16 Ibid., 404.

between his speculations and his great achievements; without the former we should never have had the latter.[17]

Again, it was Kepler's Pythagorean-Platonic bent and his deep conviction in an archetypal cosmos that led to his three laws. Moreover, his commitment to the empirical data was above reproach. Far from hampering the advancement of knowledge, his metaphysics facilitated his discoveries long before the existence of the empirical tools required for a conventional method.

Kepler's Harmonization of Christian Theism and Natural Philosophy

There are several crucial points to grasp concerning the relationship between Kepler's natural philosophy and his Christianity, and all are in some respect related to his ideas about a grand world harmony. First, he never saw naturalistic explanations as somehow removing or reducing the necessity of God in explaining the cosmos; in fact, he believed that in uncovering the workings of nature he was bringing glory to its Creator. Second, he denied that heliocentrism or the immensity of the universe in any way diminished mankind's significance; indeed he saw the absurdity in such a notion. Third, he made very clear statements about his view on the proper relationship between natural philosophy and the teachings of holy Scripture. In these ways, as in his astronomical theorizing, Kepler was a thinker well ahead of his time; though not a theologian by vocation, perhaps it can be said that he was a natural theologian in the truest and purest sense.

---

[17] J. L. E. Dreyer, *A History of Astronomy from Thales to Kepler* (New York: Dover, 1953), 410.

As previously discussed, Kepler pursued mechanistic explanations for celestial dynamics and saw compatibility between natural efficient causation and God's archetypal plan and sovereignty over the created world.[18] In other words, the idea of a mechanistic cosmos animated by a magnetic, motive force radiating from the sun (a sacramental symbol of God, the Unmoved Mover) was one more glorious example of God's wondrous design, not a diminishing of His role in creation. Rhonda Martens explains,

> "As the sun is at the center of Kepler's physical universe, God is at the center of his archetypal universe. Kepler's God was a Platonic God, an aesthetician and a geometer, who created physical things to express aesthetically pleasing geometrical constructions. As a mind, God's creation of the universe was based on ideas, ideas of quantities—shape, number, and extension—in particular."[19]

She continues:

> [Kepler] did not think the divine began at the borders of the Earthly. It belonged rather to an entirely different plane (the archetypal). Anything accessible to the senses is not divine but a material representation of it. As a result, the physical universe includes celestial bodies, and these bodies are subject to the same principles as other physical bodies, because they were all created to instantiate the archetypes. For Kepler, then, his archetypal universe permitted the exploitation of knowledge of terrestrial causes without requiring the more radical abandonment of the distinction between the divine and mundane.[20]

---

18 It should be noted, however, that Kepler's worldview was not entirely devoid of animism. See Patrick J. Boner, "Life in the Liquid Fields: Kepler, Tycho and Gilbert on the Nature of the Heavens and Earth," *History of Science* 46, no. 3 (2008).

19 Rhonda Martens, *Kepler's Philosophy and the New Astronomy* (Princeton: Princeton University Press, 2000), 48.

20 Ibid., 84.

A material world operating in law-like fashion according to natural causes was by no means one that was independent of God, the Architect behind the entire ingeniously ordered edifice. Kepler saw his planetary laws, especially his Third Law, as stunning affirmation of his Pythagorean-Platonic understanding of the universe: a world pervaded with intelligible mathematical forms and harmonies.

Kepler believed that by carrying out his investigations, the natural philosopher is gaining glimpses into the very mind of God, and by bringing this natural revelation to light for all to see he is acting as a priest in God's grand cosmic temple. In a letter to Herwart von Hohenburg dated March 26, 1598 (a very early point in Kepler's career), he wrote, "We astronomers are priests of the highest God in regard to the book of nature...I am content with the honor of having my discovery guard the doors of the sanctuary in which Copernicus performs the service at the higher altar."[21] Indeed, Kepler saw his entire life's work as one long, intensive effort to bring glory to the Maker of all things. This attitude was all the more obvious in his later years, especially in his description of his "sacred frenzy" over plundering the metaphorical vessels of the Egyptians for the furnishing of God's holy tabernacle—elucidating the harmonic law. Kepler clearly believed that the highest goal of natural philosophy was to lead men to knowledge of God.[22] In other words, it functions as natural theology.

Kepler was convinced that by studying and exegeting God's book of nature, one's worship of the Creator is vastly elevated. In the dedication letter at the beginning of the *Mysterium* he asks, "Why then as Christians should we take any less delight in [nature's] contemplation, since it is for us with true worship to honor God, to venerate him, to wonder at him? The more rightly we understand the nature and scope of what our God has founded,

---

21 Carola Baumgardt, *Johannes Kepler: Life and Letters*, 44–45.

22 Caspar, *Kepler*, 376.

the more devoted the spirit in which that is done."[23] In the dedication to the first three books of his *Epitome of Copernican Astronomy* (1619) he writes, "I have been made priest of God, the creator of the book of nature, I have composed this hymn for God the creator."[24] That he saw astronomy as holy work is further illustrated by the closing prayer at the end of the *Harmonice Mundi* (quoted here in full):

> It now remains for me, at the very last, to take my eyes and hands away from the table of proofs, lift them up to the heaven, and pray devoutly and humbly to the Father of light: O Thou who by the light of Nature movest in us the desire for the light of grace, so that by it thou mayest bring us over into the light of glory; I thank Thee, Creator Lord, because Thou hast made me delight in Thy handiwork, and I have exulted in the works of Thy hands. Lo, I have now brought to completion the work of my covenant, using all the power of the talents which Thou hast given me. I have made manifest the glory of Thy works to men who will read these demonstrations, as much as the deficiency of my mind has been able to grasp of its infinity. My intellect has been ready for the most accurate details of philosophy. If anything unworthy of Thy intentions has been put forward by me, miserable worm that I am, born and nourished in a slough of sins, which thou wouldst wish men to know, inspire me also to set it right; if I have been enticed into temerity by the wonderful splendor of Thy works, or if I have loved my own glory among men, while advancing in work destined for Thy glory, mildly and mercifully pardon it; and last, be gracious and deign to bring about that these my demonstrations may be conducive to Thy glory and to the salvation of souls, and may in no way obstruct it.[25]

---

23 Johannes Kepler, *Mysterium Cosmographicum*, 53.

24 Baumgardt, *Johannes Kepler: Life and Letters*, 122–123.

25 Johannes Kepler, *The Harmony of the World*, 491.

The final line of Kepler's prayer is indicative of the value he saw in his life's work for the purposes of natural theology. Clearly, he believed that his work pointed beyond the material realm to a creator God.

Kepler rejected the idea that the enormity of the cosmos or heliocentric cosmology suggested that mankind is less important than in the cozier, geocentric Aristotelian-Ptolemaic model. In other words, he did not see relative size or geometric location as having any bearing on human significance in the grand scheme of the world. In a letter to Herwart von Hohenburg dated December 16, 1598, Kepler quotes Copernicus, who had declared, "So great indeed is the edifice of our Almighty and Allkind Creator," and then adds that "we should feel less astonished at the huge and almost endless width of the heavens than at the smallness of us human beings, the smallness of this, our tiny ball of earth."[26] He goes on to say that man is "puny" compared with the universe, "Yet one must not infer from bigness to special importance" because if physical size indicates our significance in the eyes of the Creator, then (he quips) "the crocodile or the elephant would be closer to God's heart than man, because these animals surpass the human being in size."[27] His statement directly implies that it is absurd to equate physical size with objective value or to think that our comparative smallness is indicative of a universe that is not metaphysically anthropocentric.

As for the earth's location in orbit around the central sun, Kepler regarded this arrangement as incredibly fortunate for the natural philosopher seeking to know God's mind through its manifestation in the creation. Thus, the position of man's home planet demonstrates his privileged status in the cosmic economy. In his

---

26 Baumgardt, *Johannes Kepler: Life and Letters*, 48–49.
27 Ibid., 49.

commentary on Galileo's *Sidereus Nuncius*, Kepler writes that man is created for contemplation of the heavens, and that because he is

> adorned and equipped with eyes, he could not remain at rest in the center. On the contrary, he must make an annual journey on this boat, which is our earth, to perform his observations...There is no globe nobler or more suitable for man than the earth. For, in the first place, it is exactly in the middle of the principal globes... Above it are Mars, Jupiter, and Saturn. Within the embrace of its orbit run Venus and Mercury, while at the center the sun rotates.[28]

As historian Dennis Danielson points out, "This is clearly a complete reconceptualization of what it means to be in the center. To exercise or actualize their divine image properly, humans must be able to observe the universe from a 'central' but dynamic and changing point of view conveniently provided by what Kepler sees as this optimally placed orbiting space station of ours."[29] Kepler continues his argument about the earth's location being best suited for the work of the astronomer: "We on the earth have difficulty in seeing Mercury, the last of the principal planets, on account of the nearby, overpowering brilliance of the sun. From Jupiter or Saturn, how much less distinct will Mercury be? Hence this globe seems assigned to man with the express intent of enabling him to view all the planets."[30] Caspar offers additional pertinent details about Kepler's view:

> Was [the earth] humiliated by being pushed out of the center of the world? By no means...By its motion around the sun its

---

28 Johannes Kepler, *Kepler's Conversation with Galileo's Sidereal Messenger*, trans. Edward Rosen (Johnson Reprint Corp., 1965), 45.

29 Dennis Danielson, "The great Copernican cliche," *American Journal of Physics* 69 no. 10 (October 2001): 1032. In this article, Danielson offers an outstanding discussion of the origin of the so-called "Copernican Principle" (man's cosmic demotion) and how it has persisted in Western thought up to today.

30 Kepler, *Kepler's Conversation with Galileo's Sidereal Messenger*, 46.

> inhabitants will be enabled to ascertain the size of the world. The unchanging inclination of the earth's axis takes care of the change of seasons and brings about an equitable distribution of the sunshine on the inhabitants of the various zones...On this trip around the stationary sun, man can observe with understanding the wonder of the world in its diversity of phenomena. For everything is there because of man.[31]

Danielson sums it up well; he writes that for Kepler, "only with the abolition of geocentrism may we truly say that we occupy the best, most privileged place in the universe."[32] Truly, it seems that the Creator specifically intended the humane art of astronomy.

What came to be the most widely read of Kepler's writings were his arguments about the compatibility of heliocentrism and holy Scripture that he included in the introduction to the *Astronomia Nova* (1609)—the only portion of his work to be translated to English prior to the 1870s.[33] As aforementioned, he was essentially an accommodationist; he explains that when Scripture speaks of common, observable phenomena "concerning which it is not their purpose to instruct humanity," they "make use of what is generally acknowledged among humans, in order to weave in other things more lofty and divine."[34] In other words, Scripture accommodates its audience by speaking of the natural world according to how it appears to the average human observer. When it mentions the sun "ascending" or "descending" or the sun "standing still" (as in the book of Joshua) it is simply using language appropriate to earth-bound humanity's perspective. To further support his argument,

---

31 Caspar, *Kepler*, 386–387.

32 Danielson, "The great Copernican cliche," 1032. Kepler's argument has been adopted and expanded in contemporary intelligent design literature. See Endnote 16.

33 Johannes Kepler, *Astronomia Nova*, trans. William H. Donahue (Santa Fe, NM: Green Lion Press, 2015), 28 n. 22.

34 Ibid., 29. This remark bears a striking similarity to Galileo's later words in his 1615 letter to the Grand Duchess Christina.

Kepler cites Psalm 24, which reads "The earth is the Lord's and the fullness thereof, the world and those who dwell therein, for he has founded it upon the seas and established it upon the rivers."[35] Taken in a literalistic way, this passage seems to teach that planet earth actually floats upon the waters of seas and rivers. If someone were to suggest such a thing, Kepler says, "Would it not be correct to say to him that he should regard the Holy Spirit as a divine messenger, and refrain from wantonly dragging him into physics class?"[36] In other words, Scripture does not intend to teach the facts of natural philosophy; its statements about natural phenomena are meant to reveal fundamental theological truths. Beyond these comments, Kepler does not develop a fully fleshed-out theory on the relationship between Scripture and natural philosophy, yet he says enough to indicate his general position, which is resonant with Galileo's.

Kepler's concluding remarks about the compatibility of Scripture and heliocentrism are colored by facetious exasperation:

> But whoever is too stupid to understand astronomical science, or too weak to believe Copernicus without affecting his faith, I advise him that, having dismissed astronomical studies and having damned whatever philosophical opinions he pleases, he mind his own business and betake himself home to scratch in his own dirt patch, abandoning this wandering about the world. He should raise his eyes (his only means of vision) to this visible heaven and with his whole heart burst forth in giving thanks and praising God the Creator. He can be sure that he worships God no less than the astronomer, to whom God has granted this, that with the mind's eye he see more penetratingly, and that he himself is able and willing to celebrate his God above whatever he discovers.[37]

---

35   Psalm 24:1–2.

36   Kepler, *Astronomia Nova*, 31.

37   Ibid., 33.

Although the astronomer sees more deeply (from a philosophical perspective) into the majesty of God by studying the creation, the unlearned man may still look up to the heavens, which declare the glory of God and thus inspire equally sincere worship. The man unschooled in natural philosophy should not concern himself with the things he is ignorant of, lest he come to erroneous conclusions. The astronomer, on the other hand, should acknowledge God's wisdom made manifest in beautiful arrangements such as the motion of the earth around the sun.[38]

---

38 Ibid.

CHAPTER 8

Kepler's Tripartite Harmony

Kepler's belief that there must be unifying physical forces (natural efficient causes) and mathematical archetypes (formal causes) underpinning the world are part of his metaphysical conviction that there exists a grand cosmic harmony orchestrated by the Creator. This *tripartite harmony* (as I call it) is rooted—as seen in previous chapters—in the Christian stream of the Pythagorean-Platonic tradition; it consists of the immaterial archetype of creation grounded in the mind of God, the manifestation (material copy) of this rational, mathematical plan in the sensible world, and the intellectual powers of mankind—creatures made in the image of God.[1] Together, this harmony of archetype, copy, and image is what renders the world intelligible to the natural philosopher and aesthetically appealing to human beings in general. Rhonda Martens writes, "Kepler believed that the world was created as an expression of God's own essence...Exploring the ultimate structure of the world...amounts to a study of divine aesthetics (which would be kin to our own since we are created in God's image)."[2] Kepler was convinced that the mathematical harmonies that characterize the cosmos, such as his Third Law of planetary motion, are detectable

---

1 Kepler's self-written epitaph relates to his philosophy. See Endnote 17.

2 Rhonda Martens, "Kepler's Solution to the Problem of a Realist Celestial Mechanics," *Studies in History and Philosophy of Science* 30, no. 3 (1999): 386.

because the human mind is designed with innate paradigms of such harmonies. Thus, the archetype in the mind of God, the material copy, and the noetic powers of mankind form a three-part harmony that (as Max Caspar puts it) "seized [Kepler] most vehemently and which influenced him throughout his whole life."[3]

In an April 1599 letter to Herwart von Hohenburg, Kepler expresses his rapture over the tripartite harmony in what are perhaps his most oft-quoted (usually paraphrased) words:

> To God there are, in the whole material world, material laws, figures and relations of special excellency of the most appropriate order...Those laws are within the grasp of the human mind; God wanted us to recognize them by creating us after his own image so that we could share in his own thoughts. For what is there in the human mind besides figures and magnitudes? It is only these which we can apprehend in the right way, and if piety allows us to say so, our understanding is in this respect of the same kind as the divine, at least as far as we are able to grasp something of it in our mortal life...for the divine counsels are impenetrable, but not his material creation.[4]

The study of creation allows us (creatures with an analogous form of rationality) to think God's thoughts—to gain a glimpse into the divine archetypal plan. In a similar vein, in the introductory dedication to his *Mysterium*, Kepler writes that "God, like a human architect, approached the founding of the world according to order and rule and measured everything in such a manner, that one might think not art took nature for an example but God Himself, in the course of His creation took the art of man as an example."[5] In a letter to his former astronomy professor, Michael Maestlin, Kepler expounds upon the same theme: "God, who founded everything in

---

3   Max Caspar, *Kepler*, 377.

4   Carola Baumgardt, *Johannes Kepler: Life and Letters*, 50.

5   Ibid., 33–34.

the world according to the norm of quantity, also has endowed man with a mind which can comprehend these norms. For as the eye for color, the ear for musical sounds, so is the mind of man created for the perception not of any arbitrary entities, but rather of quantities."[6] Man's mind is imprinted with mathematical concepts so that he may discern God's mathematical archetypes.

Kepler did not use the term "archetype" in his *Mysterium*, but when he revised the book twenty-five years after its first publication, he added a note to Chapter XI in which he articulates what is essentially his epistemology of natural philosophy, and here he does include the term. He says that the principle of harmony, which he had already been so firmly persuaded of when the *Mysterium* was originally written,[7]

> has repaid me with interest over these 25 years—that is, that the reason why the Mathematicals are the cause of natural things...is that God the Creator had the Mathematicals with him as archetypes from eternity in their simplest divine state of abstraction, even from quantities themselves, considered in their material aspect.[8]

Similarly, in the introduction to Book I of the *Harmonice* he writes that

> finite things which are [mathematically] circumscribed and shaped can also be grasped by the mind: infinite and unbounded things, insofar as they are such, can be held in by no bonds of knowledge, which is obtained from definitions, by no bonds of

---

6 Gerald Holton, *Thematic Origins of Scientific Thought*, 68.

7 Caspar suggests that it was the idea expressed by Copernicus's pupil Rheticus, who had remarked in the *Narratio Prima* "that God had so arranged the world that a heavenly harmony in which each planet would have a particular place would be perfected by the six movable spheres," that first ignited Kepler's passion for harmony. See Caspar, *Kepler*, 377.

8 Johannes Kepler, *Mysterium Cosmographicum*, 125 n. 2. Here, Kepler apparently means formal cause.

constructions. For shapes are in the archetype prior to their being in the product, in the divine mind prior to being in creatures, differently indeed in respect of their subject, but the same in the form of their essence.[9]

The natural philosopher's ability to grasp the form of nature is made possible, in part, because it is "circumscribed and shaped"—limited and rationally structured. As seen in the discussion on Philolaus in Chapter 1, the idea that mathematical order and limitation make the world intelligible is deeply embedded in the Pythagorean-Platonic tradition.

More generally, Kepler's philosophy of nature—that it is rationally ordered in a manner resonant with the mind of man—was undoubtedly a product of Pythagorean-Platonic thought fused with Christian theism. In his *Harmonice*, he writes that the contemplation of the archetypes "is lofty, Platonic, and analogous to the Christian faith, looking towards metaphysics and the theory of the soul. For geometry…is coeternal with God, and by shining forth in the divine mind supplied patterns to God…[and] all spirits, souls, and minds are images of God the Creator."[10] He explains that "the Creator, the actual fount of geometry, who, as Plato wrote, practices eternal geometry, does not stray from his own archetype."[11] Later in the same work, he describes God as "the fountain of all wisdom, the constant advocate of order, the eternal and transcendental wellspring of geometry and harmony."[12] In his *Conversation with Galileo's Sidereal Messenger*, he writes that geometry "shines in the mind of God" and that a "share of it which has been granted

---

9   Johannes Kepler, *The Harmony of the World*, 9–10.

10   Ibid., 146.

11   Ibid., 407. This alleged saying of Plato (which is not found in the Platonic corpus) comes from Plutarch, *Convivia*, viii, 2. See page 407 n. 38.

12   Kepler, *The Harmony of the World*, 451.

to man is one of the reasons why he is in the image of God."[13] As Holton notes, Kepler associated mathematical quantity with the Deity; he held "that man's ability to discover harmonies, and therefore reality, in the chaos of events is due to a direct connection between ultimate reality, namely, God, and the mind of man."[14] In sum, the Keplerian view is that the human mind is shaped by God in such a way that it can be said to contain innate constructs that allow it to grasp what he calls "mathematicals" and to recognize intellectual harmonies.

In Book IV of the *Harmonice*, Kepler mentions the Platonic doctrine of recollection as it was explicated in the *Meno* dialogue; he writes, "Now Plato's view on mathematical things was that the human mind is in itself thoroughly informed on species of figures, and axioms and conclusions about things. However, when it seems to learn, it is merely being reminded by sensible diagrams of those things which it knows on its own account."[15] He then dismisses Aristotle's conception of the mind as a "blank sheet" that has "nothing...written on it, not even any mathematicals" and says that Aristotle (at least where this particular view is concerned) "is not to be tolerated in the Christian religion."[16] To reiterate, Kepler's understanding of the *imago Dei* is that it includes particular intellectual faculties and innate ideas which enable man to grasp the archetypal mathematical order and harmonies made manifest in the cosmos. This indeed seems compatible with—though perhaps not precisely identical with—the Platonic doctrine of recollection.

For Kepler, the intellect is grounded in—or perhaps *is*—the immaterial soul itself. In Book IV of the *Harmonice* he quotes extensively from Proclus's commentary on Euclid's *Elements*: "For the

---

13 Johannes Kepler, *Kepler's Conversation with Galileo's Sidereal Messenger*, 43.

14 Holton, *Thematic Origins of Scientific Thought*, 68.

15 Kepler, *The Harmony of the World*, 297.

16 Ibid., 298.

soul is a Mind, or a kind of Intellect, which reflects on itself in accordance with an Intellect which is prior to itself, having become an image of it and a representation or external copy of it... everything mathematical is first of all in the soul."[17] Kepler seems to understand the "Intellect which is prior to itself" as the mind of God, and the human mind as the intellect that is an "image of it."[18] It is the intellectual affinity between the mind of the Creator and the mind of man that makes possible the discursive thought required for natural philosophy. Martens explains that "Kepler was particularly impressed by Proclus's Platonic position that the mind is created with the imprint of mathematical ideas...[and] agreed with Proclus that mathematics, by revealing the structure of our thought, could, in virtue of the doctrine of emanation, reveal the structure of the world."[19] It could be said that the immaterial archetypes are inherent to minds—both human and divine—and the sense of harmony experienced by the natural philosopher comes about when an instance of order observed in the material world matches the mental archetype. In that moment of tripartite harmony, man truly thinks God's thoughts after Him. Gerald Holton writes, "This...is the final justification of Kepler's search for mathematical harmonies. The investigation of nature becomes an investigation into the thought of God, Whom we can apprehend through the language of mathematics."[20]

Kepler's philosophy of nature guided his thinking in other respects. As Rhonda Martens explains:

---

17 Kepler, *The Harmony of the World*, 301.

18 The relationship or identity Kepler draws between the intellect and soul should be noted, as this will be a central concern of Chapter 13.

19 Martens, *Kepler's Philosophy and the New Astronomy*, 34. Martens notes that Kepler's interpretation of Proclus was distinctly Christian. This is indicated by the intertextual bracket comments Kepler includes in the quoted Proclus passage.

20 Holton, *Thematic Origins of Scientific Thought*, 69.

The archetypes supported Kepler's realist stance in two ways. They justified his method as truth-linked, and they accounted for the accessibility to the human mind of nature's underlying structure. In particular, they justified his view that fruitfulness and simplicity are theoretical virtues on the basis of which to adjudicate rival hypotheses. Because God's essence is simple, and this simplicity is reflected in the material by archetypal correspondence, a theory that does not unify complex phenomena is unlikely to match the archetypes.[21]

Elsewhere, Martens notes how Kepler, as a realist, insisted that mathematical models should be accompanied by plausible physical interpretations.[22] This helps to explain why he continued to amend his polyhedral/harmonic theory according to Brahe's empirical data. Also significant is the fact that the archetypal universe served as justification for blending astronomy and physics, which was the bridge from ancient astronomy to the new astronomy. Because he conceived of physical objects as fundamentally geometrical, it made sense to use mathematical models to describe what he believed to be mechanical systems.[23] This approach to astronomy also allowed him to rule out hypotheses that could not be expressed geometrically and establish those that could, which turned out to be instrumental, particularly in his explication of the elliptical orbit of Mars.[24]

To the geometrical archetype of the cosmos—a heliocentric arrangement of the planets encompassed by the fixed stars—Kepler applied a sacramental interpretation, which he first mentions in the *Mysterium*. The sphere of fixed stars which make up the outermost circumference of the universe symbolized the *Logos*, the

---

21 Martens, *Kepler's Philosophy and the New Astronomy*, 169.

22 Martens, "Kepler's Solution to the Problem of a Realist Celestial Mechanics," 381.

23 Martens, *Kepler's Philosophy and the New Astronomy*, 169.

24 Ibid.

Son through Whom all was made; the sun at the center of all things represented God the Father from Whom all else emanates; and the intermediate space in which the planets revolve symbolized the Holy Spirit.[25] The universe itself was, for Kepler, a material signification of its Creator's trinitarian nature. Caspar elaborates:

> From the center point of the sphere, as the origin, the surface emerges by means of radiation, whereby the surrounding equal intervening space is produced of its own accord...All three—centerpoint, surface, intervening space—stand in most intimate relationship, in loveliest accord, in the best proportioned ratio to each other. Together they form a unity so that not once can one of them be thought of as missing without the whole being annihilated. So the secret, unfathomable existence of the divine Trinity mirrors itself for Kepler in the visible world. The world is a sphere, the sphere a picture of the Holy Trinity; consequently the world is the corporeal image of God.[26]

Kepler mentions this symbolism in many other places, including his *Optics*, the *Astronomia Nova*, the *Epitome*, an October 3, 1595 letter to Maestlin, and a March 28, 1605 letter to Herwart von Hohenburg.[27] Not only had the rational Creator structured the universe according to beautiful mathematical archetypes, the overall heliocentric structure is meant to communicate further truths about His nature. Holton elaborates on Kepler's theological symbolism:

> In the medieval period the "place" for God, both in Aristotelian and in neo-Platonic astronomical metaphysics, had commonly been either beyond the last celestial sphere or else all of space; for only those alternatives provided for the Deity a "place" from

---

25 Kepler, *Mysterium Cosmographicum*, 95. Kepler adopted and modified the trinitarian metaphor from Nicholas of Cusa. See *Mysterium Cosmographicum*, 237 and *The Harmony of the World*, 304 n. 27.

26 Caspar, *Kepler*, 379.

27 Kepler, *The Harmony of the World*, 304 n. 27.

which all celestial motions were equivalent. But Kepler can adopt a third possibility: in a truly heliocentric system God can be brought back into the solar system itself, so to speak, enthroned at the fixed and common reference object which coincides with the source of light and with the origin of the physical forces holding the system together.[28]

Unfortunately, Thomas Kuhn misleadingly characterizes Kepler's view as "sun worship." Kuhn writes, "Until after Copernicus' death the mathematical magic and the sun worship that are so marked in Kepler's research remained the principal points of explicit contact between Renaissance neo-Platonism and the new astronomy."[29] Kuhn, unless he means "worship" in a non-standard way, misconstrues Kepler's meaning and ignores the language that clearly indicates a sacramental perspective.

Caspar ends his definitive work on Kepler with a marvelous discussion of Kepler's archetype-copy-image harmony. He writes that Kepler "believed in the reality of the things outside us and in the possibility of being able to comprehend them in their essence, order and meaning…[he] was deeply convinced that there is an absolute truth. Admittedly our mind never can completely grasp this truth, but it is the noble task of scientific and philosophical research to draw nearer to it."[30] As a young man, Kepler had summed up the fundamental ideas behind the tripartite harmony with two sentences: *"Mundus est imago Dei corporea"* and *"Animus est imago Dei incorporea"*—"The world is the corporeal image of God" and "The soul is the incorporeal image of God."[31] Caspar writes,

---

28 Holton, *Thematic Origins of Scientific Thought*, 66.

29 Thomas Kuhn, *The Copernican Revolution*, 132.

30 Caspar, *Kepler*, 377.

31 Kepler makes these statements in a letter to Herwart von Hohenburg dated 9/10 April 1599. See Gingerich's endnotes in Caspar, *Kepler*, 430.

"God, World, Human—prototype,[32] copy, likeness: the circle of his thoughts is shut in this trinity. These ideas hold together and fasten tightly everything which presents itself to him."[33] Intellectually attractive mathematical ideas are fundamental to this tripartite harmony:

> Quantities can be compared with each other; they form relationships. Now God, by certain selection in creating the world has, so to speak, taken such relationships out of Himself; He has made order out of chaos, given form to matter, in accordance with the word of the Bible that everything is regulated by number, size, and weight. So to Kepler's contemplating eye the cosmos seems constructed like an ancient temple, a pyramid, a gothic cathedral, in the building of which the architect measured off the size according to aesthetic norms.[34]

Here Caspar gives a variation on the phrase that has been seen multiple times in preceding chapters—"measure, number, and weight"—to refer to Kepler's mathematical conception of the cosmic architecture.

The fact that major parts of Kepler's mathematical cosmology (namely, his polyhedral orbital distances and harmonic ratios) turned out to be false does nothing to undermine the assertion that a rational, mathematically-structured archetype pre-existed in the mind of the Creator. Although the polyhedral/harmonic theory collapsed, the idea of a rational harmony did not. Arguably, Kepler himself recognized this; in his unfettered enthusiasm over the harmonic law, he demonstrates that mathematical relationships uniting different aspects of nature into a coherent system are his essential objective. As much as Kepler loved the polyhedral/harmonic mod-

---

32 The translator of Caspar's work uses the term "prototype" rather than "archetype." Archetype is the better choice here, as prototype often denotes an unperfected working model.

33 Caspar, *Kepler*, 378.

34 Ibid., 377.

el, he saw the empirical data as paramount and continually tested the archetypal hypothesis against it.[35] In Chapter I of the *Mysterium*, he comments that he first became confident in the truth of Copernicanism because of its consistency with observations.[36] Thus, it seems more than plausible that he would have delighted in Newtonian physics as an affirmation of the mathematical harmony of the world rather than lamenting the downfall of his pet hypothesis.

The Tripartite Harmony & Contemporary Natural Theology

The chief focus of Kepler's natural philosophy was not to make a formal argument for the existence of God. However, he did acknowledge that a mathematical universe that could be elucidated by the human mind constituted evidence for divine creation. In the dedicatory letter of the *Mysterium* he writes, "I do not want to stress that I present important evidence of the creation of the universe—an evidence which has been denied by the philosophers," and then proceeds to speak about the resonance between the mathematically structured material world, the mind of God, and the mind of man.[37] Holton explains that "Kepler's God has done more than build the world on a mathematical model; he also specifically created man with a mind which 'carries in it concepts built on the category of quantity,' in order that man may directly communicate with the Deity" and concludes, in agreement with Caspar, that this idea was the "mainspring" of Kepler's life's work.[38]

---

35   Owen Gingerich, "Kepler Then and Now," *Perspectives on Science* 10, no. 3 (2002): 233.

36   Kepler, *Mysterium Cosmographicum*, 75.

37   Baumgardt, *Johannes Kepler: Life and Letters*, 33. Also see Kepler, *Mysterium Cosmographicum*, 53 for a somewhat different translation.

38   Holton, *Thematic Origins of Scientific Thought*, 68–69.

Part of the central thesis of this book is the claim that Kepler's tripartite harmony of archetype-copy-image remains a superior philosophical explanation for the mathematical structure of nature and the human ability to discern it. This is to harness the tripartite harmony—what could be called the *Keplerian natural theology*—to explain what Christopher Kaiser calls the "correspondence between the depths of the human psyche and the deep structures of the universe, between deep subject and deep object" that makes the scientific enterprise possible.[39] This project begins with the observation that there is a correspondence between the deep, rational structure of nature and the rational powers of the human mind—a correspondence that has made the natural sciences possible. From there, it asks: Why does this consonance between mathematics, mind, and matter exist? What are the metaphysical implications that can be drawn from these interconnections? The Keplerian answer is the harmony between the mind of the Creator, the manifestation of His architectonic plan in the material cosmos, and the mind of man who bears His image. The remainder of this book will be devoted to an elaboration of Keplerian natural theology in the contemporary conversation and the overarching argument that it is philosophically superior to naturalistic explanations of cosmic comprehensibility. Thus, it should be regarded as a viable perspective in the contemporary science and faith conversation.

As Part I has demonstrated, the understanding of the cosmos as a manifestation of a pre-existent rational plan in the mind of God that exhibits a mathematical orderliness detectable by creatures made in God's image is deeply rooted in ancient Western thought, going back to the early Pythagorean-Platonic tradition.

---

39 Christopher B. Kaiser, *Toward a Theology of Scientific Endeavour: The Descent of Science* (Burlington, VT: Ashgate Publishing, 2007), 143.

However, with Kepler, a devout Christian, we see a special emphasis on this idea coupled with revolutionary advancements in astronomy. This seems to constitute adequate justification for considering him the proper namesake for this unique line of argumentation in the ongoing project of natural theology.

In Part II, each conceptual element of Keplerian natural theology—archetype, copy, and image—will be examined in light of contemporary scholarship in the relevant disciplines. As will be seen, there are compelling reasons to regard this explanatory framework as more robust than ever before. It provides an intellectually satisfying, theologically orthodox, and philosophically coherent answer to the questions: Why is the cosmos intellectually transparent to the human mind? And what does this accessibility say about both the nature of the material world and man's place in it?

Part II

Archetype, Copy & Image:
The Tripartite Harmony of Kepler's Natural Theology

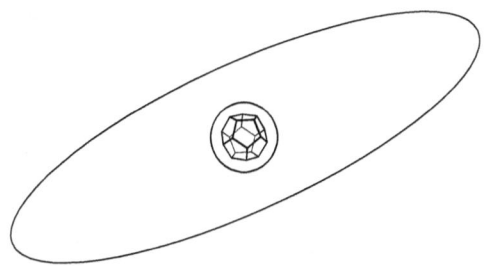

Chapter 9

Archetype—God's Mathematical Plan for Creation

As previously established, the idea of an immaterial, rational plan for the natural world goes back to the early Pythagorean-Platonic tradition, and was later embraced by monotheists who believed that the blueprint for creation pre-existed in the mind of God rather than a self-existent realm sometimes referred to as the "Platonic heaven." Keeping with the traditional Christian position, Kepler regarded geometrical and other mathematical ideas—some of which are made manifest in nature—as elements of God's mental life. In this respect at least, Kepler was not a thoroughgoing Platonist; he understood the mathematical archetype of creation as a collection of ideas, but one that is ontologically identical with a set of divine thoughts.[1]

In contemporary Christian philosophy, there is ongoing debate concerning the proper way to understand God's relationship to abstract objects—the metaphysical category that includes things like numbers, geometrical shapes, and mathematical propositions. Abstract objects *seem* to exist necessarily; it is unclear how they could *not* exist and in what sense they could be said to be created. For the traditional Christian theist, the problem is how to explain

---

1 Here and going forward, I use the terms "Platonist" and "Platonism" in the manner that they are used in the contemporary debate about God and abstract objects. This avoids interpretive issues with the original Platonic source material.

abstract objects' dependence upon God for their existence. If Keplerian natural theology (the tripartite harmony of archetype-copy-image) is to be regarded as a viable contemporary approach to explaining the mysterious resonance between mathematics, nature, and the human mind, then Kepler's understanding of mathematical entities and their relationship to the Creator (mathematical archetypes pre-existing in the mind of God) must be theologically and philosophically defensible.

Kepler's Understanding of the Archetype

Recall that Kepler was heavily influenced in his thought on the rational structure of creation by the Christian Platonism that had emerged within the Pythagorean-Platonic tradition long before his time. Some of his language echoes that of Philo of Alexandria and Augustine; in his *Harmonice* he writes that "the World, which had no previous existence, was created by God in weight, measure, and number, that is in accordance with ideas coeternal with Him."[2] Here he affirms that God's blueprint for the cosmos was *mathematical* and that mathematical ideas have eternal existence. Elsewhere in the *Harmonice*, he writes that the "quantities which have shapes" (geometrical archetypes) are intellectual in essence:

> For shapes are in the archetype prior to their being in the product, in the divine mind prior to being in creatures, differently indeed in respect of their subject, but the same in the form of their essence. Therefore in quantities shape is a kind of mental essence of them, or understanding is their essential distinguishing feature…quantities which have shapes have an intellectual essence.[3]

---

2 Johannes Kepler, *The Harmony of the World*, 115.

3 Ibid., 9–10.

Notice that Kepler specifies that the geometrical archetypes are entities in the divine intellect, which implies a realist understanding of mathematical objects. In the *Mysterium* he writes that "God the Creator had the Mathematicals with him as archetypes from eternity in their simplest divine state of abstraction, even from quantities themselves, considered in their material aspect."[4] Perhaps by "divine state of abstraction" Kepler means the perfect, immaterial state of God's mathematical ideas. Mathematical objects are eternal yet somehow ontologically dependent upon the mind of God, "the Creator, the actual fount of geometry, who, as Plato wrote, practices eternal geometry."[5] He describes God as "the fountain of all wisdom, the constant advocate of order, the eternal and transcendental wellspring of geometry and harmony."[6] Undoubtedly, Kepler would have rejected the idea that mathematics is merely a human invention or simply a figurative way of describing natural phenomena.

To reiterate, Kepler certainly believed that God was somehow ultimately responsible for the existence of the archetypal model for creation; he regarded mathematical entities as a subset of the divine mental content, some of which has been made manifest in nature. As will be seen, his view harmonizes well with more than one contemporary position. However, not all Christian analytic philosophers would agree with Kepler; some hold a strictly Platonist view, in which abstract objects such as mathematical ideas are self-existent, while others reject a realist perspective altogether and understand mathematical concepts as useful fictions. The trick is to determine which views make sense of abstract objects without compromising the traditional, biblical understanding of God and are also compatible with Keplerian natural theology.

---

4 Johannes Kepler, *Mysterium Cosmographicum*, 125 n. 2.

5 Kepler, *The Harmony of the World*, 407. This alleged saying of Plato (which is not found in the Platonic corpus) comes from Plutarch, *Convivia*, viii, 2. See page 407 n. 38.

6 Ibid., 451.

A Preliminary: Divine Aseity

According to traditional theism, God is a necessary being—that is, He is not dependent upon anything else for His existence. This is known as the divine attribute of *aseity*—from the Latin phrase *a se* ("of itself" or "from itself").[7] Aseity belongs to God alone; everything aside from Him exists *ab alio* ("through another").[8] In addition, God is understood as a personal being with supreme sovereignty over all of reality; as philosopher Paul Gould succinctly puts it, "all reality distinct from God is dependent on God's creative and sustaining activity."[9] Gould articulates what he calls the aseity-sovereignty doctrine (AD), which is comprised of two propositions:

> AD: (i) God does not depend on anything distinct from himself for his existing, and (ii) everything distinct from God depends on God's creative activity for its existing.[10]

According to AD, God is what Oxford philosopher Brian Leftow calls "the sole ultimate reality."[11] Unlike God, all other things in some way depend upon something else for their existence. As the early Church Father Irenaeus writes, "…in all things God has the pre-eminence, who alone is uncreated, the first of all things, and the primary cause of the existence of all, while all other things

---

7   William Lane Craig, *God Over All* (New York: Oxford University Press, 2016), 1.

8   J. P. Moreland and William Lane Craig, *Philosophical Foundations for a Christian Worldview* (Downers Grove, IL: IVP Academic, 2003), 504.

9   Paul Gould, "The Problem of God and Abstract Objects: A Prolegomenon," *Philosophia Christi* 13, no. 2 (2011): 256.

10 Ibid.

11 Brian Leftow, *God and Necessity* (Oxford: Oxford University Press, 2010), 27.

remain under God's subjection...For the Uncreated is perfect, that is, God."[12]

In support of AD—God's "ultimacy"—scholars point to several verses of Scripture, most often John 1:3: "All things were made through him, and without him was not any thing made that was made." In reference to the full prologue of John's gospel, William Lane Craig remarks, "Notice that in John's view, although God and His Word (*logos* in the Greek) simply *were* in the beginning (v. 1), everything else is said to have *come into being* through him (v. 3)."[13] He notes that the Greek term used in v. 3 is *ginomai*, which means "to become," "to originate," or "to be created," and explains that this sets up a contrast between God (including His Word), Who had no ultimate beginning, and the rest of the world.[14] The book of Isaiah also contains statements that seem to support AD: "for I am God, and there is no other; I am God, and there is none like me" and "Thus says the Lord, the King of Israel and his Redeemer, the Lord of hosts: 'I am the first and I am the last; besides me there is no God...'"[15] Leftow quotes these, along with several other passages, and comments:

> However exactly one parses these texts, if they make claims about God's nature at all, they assert uniqueness in respect of being 'first,' 'last,' greatest, and divine, and also some sort of uniqueness *tout court* ('none like me'). Some of these are respects of ultimacy. To be last just is to be ultimate, or perhaps ultimate in duration—that which will outlast any other thing. Being first is also a way to be ultimate—its first member is the last point to which one could

---

12   Irenaeus, *Against Heresies*, in Ante-Nicene Fathers 1, ed. Alexander Roberts and James Donaldson (Peabody, MA: Hendrickson Publishers, 2012), 521–522.

13   William Lane Craig, *God Over All*, 14.

14   Ibid.

15   Isaiah 46:9, 44:6.

trace a series back. Being greatest is being ultimate in value, the last point reached in an ascent along a value-scale.[16]

Thus, God's aseity is a great-making property; it is part of what it means to be the greatest possible being, the Supreme, the Ultimate.

The Problem of God and Abstract Objects

Intuitively, mathematical entities and propositions seem to be things that exist independently of human minds. In a universe devoid of human beings or any other rational creatures, 2 + 2 would still equal 4, and the Pythagorean theorem would still be true.[17] Some great thinkers of the Western Tradition have entertained the question of whether this means that mathematics is somehow "out there"—a transcendent, abstract reality that we discover through reasoning. Consider Augustine's perspective:

> As for the study of number, it is surely clear even to the dullest person that it was not instituted by men, but rather investigated and discovered. Virgil wanted the first vowel of *Italia*—traditionally pronounced short—to be long, and made it long, but nobody can bring it about by willing it that three threes are not nine, or that they fail to make a squared number, or that the number nine is not thrice three, or one and a half times six, or twice no number (for odd numbers are not divisible by two). So whether numbers are considered purely as numbers or used in accordance with the laws that govern figures or sounds or other kinds of motion, they have fixed rules, which were not in any way instituted

---

16   Brian Leftow, *God and Necessity*, 4.

17   The Pythagorean theorem (which is only traditionally attributed to Pythagoras) says that in Euclidean geometry, the square of the hypotenuse of a right triangle is equal to the sum of the squares of the other two sides.

by human beings but discovered by the intelligence of human brains…it is from him [God] that all things have their existence.[18]

Augustine clearly saw the science of number as something that is ontologically independent of the human mind; the grounding of its existence, he believed, is God Himself. Thus, the art of mathematics is an exercise in the rational apprehension of an immaterial, divine reality.

The belief that mathematics is discovered rather than invented has not been limited to traditional theists. For example, in his book *Principles of Mathematics*, atheist philosopher and mathematician Bertrand Russell (1872–1970) writes, "Arithmetic must be discovered in just the same sense in which Columbus discovered the West Indies, and we no more create numbers than he created the Indians."[19] Russell's contemporary, English mathematician (and non-theist) G. H. Hardy (1877–1947), also affirmed the independent existence of mathematics. In his famous work *A Mathematician's Apology* he writes:

> For me, and I suppose for most mathematicians, there is another reality, which I will call "mathematical reality"…Some hold that it is "mental" and that in some sense we construct it, others that it is outside and independent of us. A man who could give a convincing account of mathematical reality would have solved very many of the most difficult problems of metaphysics. If he could include physical reality in his account, he would have solved them all…I believe that mathematical reality lies outside us, that our function is to discover or observe it, and that the theorems which we prove, and which we describe grandiloquently as our "creations," are simply our notes of our observations.[20]

---

18 Augustine, *On Christian Teaching*, 62–63.

19 Bertrand Russell, *Principles of Mathematics* I (London: Cambridge University Press, 1903), 451.

20 G. H. Hardy, *A Mathematician's Apology* (Cambridge: Cambridge University Press, 2012), 123–124.

Russell and Hardy, like Augustine, clearly affirmed that mathematical truths are independent of human minds yet somehow accessible through reason.

Still today, there is no consensus among non-theists about how to explain the existence of abstract objects. Like Russell and Hardy, some are Platonists who regard things like numbers, geometrical shapes, and mathematical functions as brute facts that have no ontological grounding, while others insist that mathematics is merely a human invention, an ever-expanding edifice containing a variety of systems that allow us to talk meaningfully and consistently about patterns and logical relationships we encounter in the world. As will be discussed in later chapters, some prominent non-theist physicists, both past and present, have taken a hard Platonist view of mathematical objects at least in part because of the astonishing precision with which mathematics maps onto the material world, a lock-and-key fit that has facilitated scientific discovery in countless ways. For these Platonists, the fact that highly complex mathematical systems can model the fundamental structure of physical reality seems to suggest that there are mind-independent mathematical structures that in some sense serve as an underlying rational pattern for the material world. Yet, as non-theists, they are unwilling to go one step further by affirming that a transcendent mind is the grounding for mathematical truths.

The question for present purposes is whether the existence of abstract objects in the realist sense is compatible with the orthodox Christian doctrine of divine aseity. As previously discussed, Kepler understood the mathematical archetypes of creation as eternal ideas in the mind of God. However, the current debate about God's relationship to abstract objects is broader than mathematical concepts; it also includes things like properties (such as blueness, roundness, omniscience) and propositions (statements that can be true or false). Although Kepler did not concern himself with these, the relevant contemporary literature tends to lump the different species of abstracta together. This often creates confusion, since some arguments for the different views and critiques of competing views do not seem to apply to every type of abstract object, such as Kepler's mathematical archetypes. Nevertheless a sampling of these arguments and critiques will be included for the sake of an accurate portrayal of the contemporary debate.

The remainder of this chapter will be devoted to an examination of the most extreme realist view, theistic Platonism. Then, the realist and non-realist alternatives will be analyzed in Chapter 10.

Theistic Platonism

As previously explained, AD (the doctrine of God's aseity and sovereignty) says that God is the only uncreated being and that everything distinct from God depends upon His creative activity for its existence.[21] God alone is ultimate. However, some theists claim that there are uncreated abstract objects (the "Platonic horde" that includes, among other entities, numbers and their mathematical relations) that exist in the same sense that concrete things exist—a view commonly referred to in the contemporary scholarly discussion as *Platonism*. As Alvin Plantinga puts it, "Platonism with respect to these objects is the position that they do exist...in such a way as to be independent of mind; even if there were no minds at all, they would still exist."[22] Christopher Menzel explains that in Platonism, "most, if not all, abstract objects are thought to exist necessarily."[23] The term *heavyweight Platonism* is used to distinguish

---

21 Paul Gould, "The Problem of God and Abstract Objects: A Prolegomenon," 256.

22 Alvin Plantinga, *Where the Conflict Really Lies: Science, Religion, and Naturalism* (New York: Oxford University Press, 2011), 288.

23 Christopher Menzel, "Theism, Platonism, and the Metaphysics of Mathematics," *Faith and Philosophy* 4, no. 4 (1987): 365.

this view from so-called "lightweight" Platonism, which sees abstract objects as merely semantic objects.[24] In what follows, the term *Platonism* should be understood as referring to the heavyweight form. Mathematical objects will receive some emphasis in this discussion due to their direct relevance to the Keplerian idea of a mathematical archetype for creation pre-existing from eternity in the mind of God.

Many (but not all) advocates of theistic Platonism find this view persuasive because of something called the Quinean Indispensability Argument, named for its twentieth-century originator, W. V. O. Quine. There are multiple revised ("neo-Quinean") versions of this argument in the contemporary discussion, but generally speaking, they can be summed up as follows:

I. If a simple sentence, such as "x is Y" or "x is R-related to q" is literally true, then the objects that its singular terms refer to exist. Also, if an existential sentence, such as "there is an x," is literally true, then the object it denotes does exist.

II. There are literally true simple sentences containing singular terms that could only refer to things that are abstract objects. Also, there are literally true existential sentences whose existential quantifiers range over things that could only be abstract objects.

III. Therefore, abstract objects exist.[25]

The first part of premise I says that ontological commitments are made by the use of singular terms—words and phrases that are used to single out a thing (represented in premise I by lowercase letters). These include demonstrative terms such as "this goose" and "that bridge"; proper names such as "Elvis Presley" and "Empire State Building"; and definite descriptions such as "the sword in

---

24 See discussion in Craig, *God Over All*, 8–12.

25 Ibid., 45–46.

the stone" and "the kayak on the beach." Premise I specifies that if a simple sentence containing a singular term is literally true, then whatever is denoted by the singular term exists. For example, consider the sentence: "The Empire State Building is a skyscraper." According to the neo-Quinean argument, the literal truth of this sentence entails that the Empire State Building actually exists. Premise I also claims that ontological commitments are made by using literally true existential sentences. For example, "There is a penguin in Antarctica" is literally true if at least one penguin lives in Antarctica. Thus, the literal truth of this sentence commits one to the existence of a penguin. Premise II claims that literally true simple sentences may have singular terms that are not concrete objects. An example would be "$2 + 5 = 7$." The referents of the terms in this number sentence are abstract objects, and as Craig notes, premise II "excludes taking such mathematical discourse to be some sort of figurative language, not to be taken literally. It claims that at least some abstract discourse is literally true and therefore commits its user to the reality of abstract objects."[26] The second part of premise II says that literally true existential statements such as "There is a prime number between 6 and 8" commit one to the existence of that prime number. The neo-Quinean indispensability argument is controversial, and there is a wide range of responses in contemporary literature. For present purposes, it is not necessary to explore these responses; the point here is that the neo-Quinean argument is one way some (but not all) Platonists defend their position, and the rejection of either premise I or II of the argument is one strategy of anti-Platonists.

In a helpful essay entitled "Did God Create Shapes?" Notre Dame philosopher Peter van Inwagen, a prominent proponent of theistic Platonism, defends his view using an example that is appropriate within the context of this project—the geometrical

---

26 Ibid., 50.

set of Platonic polyhedra.[27] Van Inwagen begins by pointing out that there are shapes that are nowhere physically instantiated yet are possible (could be instantiated), and that accepting this fact seems to commit us to the Platonic existence of possible shapes.[28] He offers a thought experiment in support of this claim. Imagine, he says, that in the third century BC a critic tells Euclid—who devised a proof for the five Platonic polyhedra—that the icosahedron does not exist because there are no icosahedral objects (that they knew of in that historical era). In other words, the critic says that there cannot be a three-dimensional shape that nothing actually has. However, Euclid's proof indicates that the icosahedron is a geometrical reality in the abstract sense, and a craftsman *could* construct an icosahedral object in order to physically demonstrate that Euclid's proof is correct.[29] Van Inwagen suggests restating the conclusion of Euclid's proof accordingly: "It is *possible* for there to be five distinct physical objects each of which is Platonic and each of which is shaped differently from each of the others; it is *not* possible for there to be *six* distinct physical objects each of which is Platonic and each of which is shaped differently from each of the others."[30] True statements about what is mathematically possible in terms of objects with a Platonic shape seem to imply that there really *are* such things as the Platonic shapes even if there were no physical objects exhibiting them; they *could* be constructed in tangible form based upon the parameters in Euclid's proof.

Van Inwagen goes further: "What I have said implies that a shape can exist if nothing has it. But there is more: a shape not only *can* exist if nothing has it, it *must* exist…Every shape must exist whether anything has it or not, for every shape is *necessarily* ex-

---

27 The five-member set that contains the regular convex polyhedra, each one having faces all of the same size and shape, and edges of the same length.

28 Peter van Inwagen, "Did God Create Shapes?" *Philosophia Christi* 17, no. 2 (2015): 285.

29 Ibid., 286–287.

30 Ibid., 286.

istent."[31] Consider the fact that if God has absolute freedom in His creation, He could have chosen not to create anything at all. Yet, even in that case, argues van Inwagen, God would *know* about the geometrical concepts of Platonic solids and would know of them as possible shapes for created objects: "God knows about cubical things as possibilities. That is, he knows that he has the power to create a universe some of whose constituent objects are cubes. So he must know about the shape 'the cube'...And of course, if God contemplates a shape, there is a shape that he is contemplating."[32] In addition, God would know that it is impossible to add a sixth member to the five-member set known as the Platonic polyhedra.

In sum, to say that there are possible shapes that nothing actually has seems to imply the real existence of abstract objects. Van Inwagen claims to be comfortable with the coexistence of God and necessarily existing abstract objects for essentially the same reason that he is comfortable with affirming that there is no conflict between Matthew 19:26 (which says that for God everything is possible) and the statement that it is impossible for God to do anything that involves a logical contradiction (such as creating a square circle) or going against His own nature (by breaking His own promise, for example).[33] To say that God does not have the freedom to create more Platonic polyhedra is on par with saying that He does not have the freedom to create a rock so heavy He cannot lift it; abstract objects are simply not the kinds of things that *could* be created, as they cannot enter into any sort of causal relation.[34] "In my view," van Inwagen explains, "when we say that God is the creator of all things...I don't think that we mean—at any rate that we have to mean—that he is the creator of abstract

---

31   Ibid., 288.

32   Ibid., 288–289.

33   Ibid., 289.

34   Peter van Inwagen, "God and Other Uncreated Things," in *Metaphysics and God: Essays in Honor of Eleonore Stump*, ed. Kevin Timpe (London: Routledge, 2009), 5.

objects, of things like propositions and attributes and numbers… and shapes."[35] Ultimately, van Inwagen accepts the existence of uncreated, necessarily existing abstract objects but denies that they pose a fatal problem for an orthodox doctrine of creation.

For van Inwagen and other theistic Platonists then, God's ultimacy is not truly compromised by the existence of ontologically independent abstract objects, such as mathematical objects, any more than it is affected by logical impossibilities. However, some scholars disagree. According to Menzel, if Platonism is true, "rather than the sovereign creator and lord of all things visible and invisible, God turns out to be just one more entity among many in a vast constellation of necessary beings existing independently of his creative power."[36] In a similar vein, Craig argues:

> Any object, whether concrete or abstract, which is uncreated will fatally compromise God's being the sole ultimate reality. The theist can happily admit the existence of created, contingent, transitory abstract objects like the Equator or Beethoven's Fifth. What he cannot allow is the existence of things which are as ontologically ultimate as God. The reason abstract objects are at the center of this controversy is simply because they are the most—perhaps only—plausible candidates for uncreated, necessary, eternal objects apart from God Himself.[37]

The distinction Craig makes between contingent and necessary abstract objects should be noted, as that represents a key point in the debate. It is the alleged abstract objects that exist *necessarily* that pose the problem. As Leftow puts it, "If the eternal truths are ultimate, it is as if we stand with God under a sky not of His devising."[38] Indeed, the idea that there could be any reality at all in the

---

35 Van Inwagen, "Did God Create Shapes?", 290.

36 Menzel, "Theism, Platonism, and the Metaphysics of Mathematics," 365.

37 Craig, *God Over All*, 6.

38 Leftow, *God and Necessity*, 72. Here, Leftow is describing Descartes's contention.

absence of God or that there are self-existent objects that dictate limitations upon God's creative activity is theologically problematic, to say the least. At the same time, it seems contradictory to suggest that mathematical objects, which seem to exist necessarily, are somehow the result of God's creative activity.

Chapter 10

Three Main Alternatives to Theistic Platonism

The task that presents itself to those who reject theistic Platonism is to explain how things like mathematical objects could exist necessarily and still be, in some sense, ontologically dependent upon God. At first blush, this seems like an impasse, but there are various philosophical strategies that attempt to untie this alleged Gordian knot. Generally speaking, the different approaches fall into one of two main categories: realism and anti-realism. As these terms imply, the realist sees mathematical objects as entities that truly exist in some fashion, while the anti-realist does not. There are quite a few subcategories that fall under each category, and full explanations of their individual nuances is a tedious endeavor due to a lack of uniformity in terminology and discrepancies in the philosophical characterizations of various views.[1] Fortunately, explicating each and every view is unnecessary for our present purposes. It will be sufficient to consider the three main options and determine which one (or more) aligns with Kepler's. If a reasonable defense can be offered for that view (or views, as the case may be), then the first string of Keplerian natural theology is in good shape.

---

1 For a useful classification chart, see William Lane Craig, *God Over All*, xii.

Option 1: Absolute Creationism

One alternative realist view is absolute creationism, according to which abstract objects are creatures; that is, wherever or whenever they exist, they exist as the result of God's creative activity.[2] This view is sometimes referred to as a modified Platonism, because it maintains that at least some abstract objects exist in a very real sense—they have objective ontological status. In their seminal paper entitled "Absolute Creation,"[3] Thomas Morris and Christopher Menzel attempt to reconcile the existence of necessarily existent abstract objects with the doctrine that all things have creaturely dependence upon God.[4] What they seek is an asymmetrical ontological dependence between abstract objects and God that preserves God's status as the ultimate reality and source of all things. If the entire assemblage of necessary truths (such as those of logic) and necessary objects (such as numbers) are thought of as the "framework of reality" that serves as the structure for both actualized reality and all possible realities, then (Morris and Menzel argue) the main question to ask is "whether, in addition to holding God responsible for the existence of every contingent reality that is structured by the framework, it can be intelligible and coherent for theists to hold God responsible for the framework itself."[5] In other words, can the claim that God is the creator of *all* things

---

2   Craig, *God Over All*, 55–56.

3   It must be noted that there has been unfortunate ambiguity in terminology over the last several decades, partly due to the fact that Thomas Morris and Christopher Menzel to some extent conflate Absolute Creationism with Theistic Conceptualism, which is a *non*-Platonic view. Here, every effort will be made to carefully delineate the two separate views according to the contemporary nuances. For a full discussion of the conceptual clarity problem in this debate, see William Lane Craig, "Absolute Creationism and Divine Conceptualism: A Call for Conceptual Clarity," *Philosophia Christi* 19 no. 2 (2017): 431–438.

4   Thomas Morris and Christopher Menzel, "Absolute Creation," *American Philosophical Quarterly* 23, no. 4 (October 1986): 353.

5   Ibid.

apart from Himself be held consistently with the contemporary Platonist view of abstract objects? They say yes and propose absolute creationism as a viable solution.

Morris and Menzel suggest that all properties (the category into which they place numbers) and relations are the result of divine intellection, a productive activity upon which such objects depend for their existence.[6] Propositions—such as those contained in Euclid's proof for the five Platonic polyhedra—can be constructed from properties (God's fundamental concepts) through divine conceiving.[7] Put more simply, God's concepts (properties and relations) are the building blocks of His thoughts (propositions). Morris and Menzel use the proposition 2 + 2 = 4 as an example in their explanation for how mathematics fits into this model: "The number 2, the number 4, the relation of addition, and that of equality are all divine concepts, all products of the divine conceiving activity. The existence of the proposition that 2 + 2 = 4 is thus the existence of a divine thought. Its [necessary] truth is also a function of that divine conceiving activity."[8] God is responsible (by His intellective activity) for the essential nature of "twoness" which includes the fact that when it is added to itself the result is 4. The mathematical proposition 2 + 2 = 4 is necessarily true and therefore part of the framework of reality for this world and every possible world. Thus, "God can be held to be not just a delimiter of possible worlds, but the absolute creator of such worlds, responsible for the abstract existence and intrinsic features of the entire framework of reality."[9] According to absolute creationism conceived in this way (what Morris and Menzel call "the-

---

6 Ibid., 355.

7 Ibid.

8 Ibid., 355–356.

9 Ibid., 356.

istic activism"), God is responsible for the creation of necessarily existing abstract objects as well as their necessity.

Absolute creationism is vulnerable to something called the *bootstrapping objection*. The idea is that in order for God to be a creator in the first place, He would have to possess certain properties, but properties themselves are (according to some) abstract objects. For example, for God to create anything at all, He would have to already have the property of *being powerful*. There seems to be a vicious circularity that results—if the property of being powerful had to exist logically prior to any creative activity, then how is it that God created the property of being powerful?

To overcome the bootstrapping objection, Paul Gould defends a version of absolute creationism he calls Modified Theistic Activism (MTA).[10] According to MTA, it is possible for God to will the existence of abstracta, even abstracta that are wholly separate from His being (existing in the Platonic heaven). Gould argues that "God's essential properties exist as uncreated constituents of the divine substance whereas God's thoughts and ideas (propositions and concepts) are created by God via divine intellectual activity."[11] God's essential properties *qua* divine properties are exempted from His creation since He is "a fundamental whole that is metaphysically prior to its properties, parts, and powers."[12] This is the modification included in MTA. Clearly, Gould disagrees with van Inwagen's assertion that abstract objects are not the kinds of things that could be *caused* to exist.

Keith Yandell has objected to MTA; he argues that if God is said to be responsible for the existence of a necessarily existing object (N) that, by definition, cannot *not* exist and has existed "eternally timeless or everlasting," then "it cannot be the case that

---

10 Paul Gould, "In Defense of Christian Platonism," unpublished manuscript.

11 Ibid.

12 Ibid.

it was possible that God not create N, for then it would be possible that N not exist."[13] Yandell believes that this results in a divine emanationism in which God's existence entails the existence of all necessarily existing objects. Yandell says,

> God has no choice but to create something (the abstracta) distinct from God...Libertarian freedom to create does not belong to God, at least regarding God's inevitable impersonal cohorts. At least often, if a particular sort of abstract object exists—numbers, sets, propositions, properties—there are an infinity of them. If one simply means by free creation by God that nothing external to God brings God to create—nothing not part of God's nature moves God to create anything—then that sort of freedom exists even though God by nature cannot not emanate abstracta. Roughly, God has only compatibilist freedom regarding creation of at least abstract objects. But it is a freedom that removes any chance that God not be emanatatively active.[14]

If Yandell's objection is right, then MTA does not fully preserve God's creative sovereignty. The key question is whether it is unreasonable to suppose that one necessary being can cause another—how to make sense of the *free creation* of a necessarily-existing object.

Gould has offered a possible solution by suggesting that divine creation can be thought of as having three moments in logical (rather than temporal) succession. The first logical moment he calls the Biggest Bang, in which "God freely, spontaneously, and eternally thinks up all possible creatures and all possible states of affairs...all possible individuals and possible worlds are set—in virtue of God's intellectual activity."[15] Like Morris and Menzel,

---

13 Keith Yandell, "God and Propositions," in *Beyond the Control of God?*, ed. Paul Gould (New York: Bloomsbury Academic, 2014), 26.

14 Ibid., 31.

15 Paul Gould, "Theistic Activism and the Doctrine of Creation," 291. Gould borrows this general idea from Brian Leftow. See Leftow, *God and Necessity*, 272–273.

Gould affirms that concepts are divine ideas and propositions that can be construed as divine thoughts.[16] It is reasonable, he argues, to understand God's intellection as the creative act, since "the relation between a thought and a thinker is most naturally understood as a *productive* relation."[17] Significantly, he specifies that the creation of possibilia is not determined by the content of the divine nature; God freely and spontaneously invents possibilia, creating their natures as well as determining their modality—and does so from eternity.[18] In the second logical moment of creation, the Bigger Bang, God creates a Platonic horde of properties and relations related to the construction of any actual universe He creates.[19] In Keplerian terms, this constitutes the archetype for creation. The third logical moment is the Big Bang, the point at which the temporal physical universe comes into being.

According to absolute creationism, in which the Platonic horde is ontologically dependent upon God's creative activity, there is another potential problem related to God's creative freedom. Suppose that God had chosen not to create anything at all. It would still be the case that the concepts of "oneness" and "threeness" would automatically exist, because God is *one* being with a *trinitarian* nature, which seems to entail the reality of one and three. The problem is, as William Lane Craig has pointed out, if even one numerical concept exists, then all other numbers exist by a sort of ontological default. Craig writes, "Even if God should exist timelessly in the absence of a created world of concrete objects, it would nonetheless be the case that the number of concrete objects is 1 (namely, God). So the number 1 would exist. But if the number 1 exists, all the rest of the natural numbers

---

16 Gould, "Theistic Activism and the Doctrine of Creation," 291.

17 Ibid.

18 Ibid.

19 Ibid.

generated by the successor relation also exist."[20] Thus, according to absolute creationism, God could not have chosen *not* to create numbers or, for that matter, any other abstract objects related to His nature and being. "Absolute creationism," says Craig, "robs God of His freedom with respect to creating."[21] Necessarily existing abstract objects such as numbers and their relations turn out to be non-voluntaristic consequences of God's existence. On the surface, this seems to compromise the traditional, orthodox understanding of creation as something over which God has comprehensive sovereignty. However, Gould and others remain convinced that, with further philosophical work, all abstract objects "can be safely brought either into the mind of God or located in Plato's heaven, without violating God's aseity or sovereignty."[22]

Option 2: Theistic Conceptualism

Another realist alternative to theistic Platonism is what Greg Welty has called theistic conceptual realism, also known as theistic conceptualism.[23] Theistic conceptualism is, in contemporary terms, anti-Platonist; it denies the existence of a horde of abstract objects existing in a Platonic heaven. As Welty construes it, theistic conceptualism identifies abstract objects with God's thoughts; it "holds that [abstract objects] are necessarily existing, uncreated divine ideas that are distinct from God and dependent upon God."[24] It is important to note that some, such as Craig, define theistic

---

20   Craig, *God Over All*, 57.

21   Ibid., 58.

22   Paul Gould and Richard Davis, "Modified Theistic Activism," in *Beyond the Control of God?*, 62.

23   Greg Welty, "Theistic Conceptual Realism," in Gould, *Beyond the Control of God?*, 81.

24   Ibid.

conceptualism a bit differently by saying that, in this view, abstract objects do not exist as such. Rather, God's thoughts, which are *concrete* in the philosophical sense, do the *work* of abstract objects.[25] Yandell agrees with this description; he says that "God's ideas or mental states are not abstract, and hence are not abstract objects. TCR [theistic conceptual realism], a perfectly fine view for one to take, cannot really mean that there are abstract objects, and that they are ideas in the mind of God."[26] According to either definition of theistic conceptualism, the entities in question have real existence within the divine mind. Alvin Plantinga (a theistic conceptualist) explains: "According to classical versions of theism, sets, numbers, and the like…are best conceived as divine thoughts. But then they stand to God in the relation in which a thought stands to a thinker. This is presumably a productive relation: the thinker produces thoughts."[27] Plantinga insists that these thoughts, as part of God's mental life and thus grounded in His essential rationality, exist eternally and necessarily—as abstract objects.

Theistic conceptualism is an attractive view in that it claims to neatly preserve the doctrine of divine aseity while recognizing the timeless, objective reality of things like mathematical objects, mathematical truths, and possibilia. Plantinga says, "It is… extremely tempting to think of abstract objects as ontologically dependent upon mental or intellectual activity in such a way that either they just are thoughts, or else at any rate couldn't exist if not thought of."[28] He is convinced that if an idea is necessarily true, as any mathematical truth seems to be, then it *exists* necessarily, thus

---

25   Craig, *God Over All*, 50.

26   Keith Yandell, "Response to Greg Welty," in *Beyond the Control of God?*, 97.

27   Alvin Plantinga, *Where the Conflict Really Lies*, 291.

28   Ibid., 288.

eternally.²⁹ Perhaps the only viable solution, he suggests, is that mathematical truths are thoughts in the divine mind.³⁰

Stephen Parrish has helpfully contributed to the conversation in his defense of theistic conceptualism. For one thing, he argues that even if the entities in question are concrete thoughts in the mind of God, human beings come to grasp them through a process of *abstracting* ideas from the particular things that constitute the created world.³¹ Thus, he says, it seems appropriate to think of things like numbers and mathematical propositions as *concrete* ideas in the mind of God but as *abstract* relative to human minds.³² Parrish also suggests that immaterial information serves as a neutral intermediary between the ideas in the mind of God and their manifestation in the physical world.³³ For example, according to theistic conceptualism, the concept of sphericity is grounded in the divine rationality, but this concrete thought cannot be said to also be *in* the spherical orange. Rather, says Parrish, the object (here, the orange) exemplifies the *informational content* of God's thought.³⁴ The rational human observer of the orange abstracts the information about sphericity from the visual and tactile experience of the orange. "The world is modeled on, or created in, the image of the ideal objects that exist in the mind of God," writes Parrish.³⁵

Parrish concludes that perhaps the best way to characterize things like geometrical shapes and mathematical objects is as *necessarily thinkable things* that have certain definite properties.³⁶ They are

---

29   Ibid.

30   Ibid.

31   Stephen E. Parrish, "Defending Theistic Conceptualism," *Philosophia Christi* 20, no. 1 (2018): 102.

32   Ibid. Parrish credits Angus Menuge for this perspective.

33   Ibid., 107.

34   Ibid.

35   Ibid.

36   Ibid., 110.

not separable things as abstract objects are in Platonism, because they are grounded in the mind of God, which gives them eternality and necessity. "Since God thinks everything at once," says Parrish, "they are all part of one grand thinking of everything that can be, or even cannot be, that is part of God's nature to think."[37] Some of God's thoughts are objects from which the things of the universe are copied, functioning as archetypal models for the structure of sensible reality (the first string of Keplerian natural theology).[38] Welty explains that "the divine ideas constitute all possible blueprints for any act of creation."[39] He insists that God's thoughts are not *creatures*; they are not things that God has *produced* from eternity. In his older literature, Plantinga concurs on this point, based upon the conflict between the plain understanding of the term "create" and the modality of necessity; he says that "a thing is created only if there is a time before which it does not exist" and that things like numbers and necessarily true mathematical propositions "have no beginnings."[40] "Necessary states of affairs," says Plantinga, "do not owe their actuality to the creative activity of God."[41] As an example of a necessary state of affairs, he offers "7 + 5 equaling 12."[42] As aforementioned, in more recent literature, Plantinga regards thoughts—including God's—as being *produced* by the thinker who has them, so perhaps his view concerning creative activity and necessity evolved over the years in between.

Craig has voiced concern about how, according to theistic conceptualism, to understand *where* God's thoughts *are* and what that "location" entails about the nature of their ontological rela-

---

37 Parrish, "Defending Theistic Conceptualism," 111.
38 Ibid.
39 Welty, "Theistic Conceptual Realism," 93.
40 Plantinga, *The Nature of Necessity* (New York: Oxford University Press, 1982), 169.
41 Ibid.
42 Ibid.

tionship to God. It seems natural, however, to understand God's mental activity as part of His mind, and thus grounded in His nature. Craig concedes this; he says, "I think that we have to admit that there remains intuitively a sense, difficult to articulate, in which divine thoughts existing 'inside' God do not seem to violate divine aseity as do uncreated, Platonic abstract objects existing 'outside' God."[43] However, Craig remains dissatisfied with the idea of God's thoughts being "inside" of God. Perhaps it is enough to say that God's thoughts are entirely ontologically dependent upon Him (thus His aseity is unthreatened) and that mathematical objects are part of the furniture of His mental life. In other words, they are God's concepts of which He has knowledge. Because God is essentially rational, just as He is essentially good, there seems to be no real problem in concluding that necessary objects, such as mathematics, share His modality. It could be that the world of mathematical objects, grounded in God's nature, includes things like the set of Platonic solids and the idea that $2 + 2 = 4$ in the same way that the eternal, objective moral law, grounded in God's essential character, includes the truth that killing people for fun is inherently evil. Beyond this, any further details about God's mind and its content may be permanently inscrutable. After all, traditional theism fully recognizes that there are aspects of God that are opaque to finite minds.

Another objection to theistic conceptualism raised by Craig involves what it means to say that a basic mathematical object, such as a number, exists in the fundamental sense as a thought in the mind of God. He says, "God's thought of the number 2 is about 2. But then His thought is not 2, but something distinct from 2, 2 is what He is thinking about. But He is not thinking about His

---

43 Craig, *God Over All*, 81.

thought; He is thinking about 2. Therefore, His thought cannot be 2."[44] This objection might be represented by a syllogism, as follows:

1. God's thought of 2 is about 2.

2. But if God's thought is about 2, then His thought is not 2 itself.

3. Therefore, God's thought is not identical to 2.

It seems that the second premise is the only one with potential vulnerability. Perhaps it can be objected that premise two is unjustifiably anthropomorphic—when it comes to God's mind, it could be the case that things in some sense go differently. It could be that numbers and other mathematical objects are somehow components of His rationality that can also serve as objects of His active thought. Moreover, it seems right to say that rationality itself requires concepts like numbers, so again, perhaps they are simply facets of the divine mind, which is the paradigm of rationality.

Option 3: Anti-realism

Some Christian theists, dissatisfied with the two main anti-Platonist realist options, take a more radical approach by denying that mathematical objects exist in the metaphysically heavy sense, whether as created abstract objects or as thoughts in the mind of God. Instead, they advocate for one of several anti-realist options that fall into a general category often referred to as *nominalism*.[45]

---

44 Craig, *God Over All*, 92.

45 Craig believes that the label "nominalism" can be misleading. The old problem was whether universals exist (nominalists say no), but the new problem is whether abstract objects exist (nominalists say no). See discussion in William Lane Craig, "Anti-Platonism" in Gould, *Beyond the Control of God?*, 116.

According to Craig, who endorses a figuralist form of anti-realism, making true statements about things such as numbers, geometrical shapes, and sets need not imply the existence of such things in the metaphysically heavy sense; rather, such terms may be understood as metaphorical ways of speaking about features of concrete things. For example, to say, "There are five oranges in the bowl," does not commit one to the existence of the number five; such a statement (says the nominalist) is merely a linguistic description about particulars—in this case, a collection of oranges. The same would be true for the sentence, "Adding two more oranges to the bowl makes seven oranges." Two, seven, and the relation $5 + 2 = 7$ are not real entities, says the nominalist; they are merely ways of describing the changing quantity of the oranges.

The appeal of nominalism is twofold; it offers a more economical ontology (by requiring fewer kinds of things in the real world) and while doing so preserves divine ultimacy. As Gould explains, according to nominalism, "the problem of God and abstract objects is dissolved—there are no abstract objects…God alone exists *a se* and creates all reality distinct from himself."[46] Craig expresses his favor for such a simplified ontology; he says that

> it is very puzzling that objects should fall into two so radically different and exclusive categories as abstract and concrete. It would be much more appealing to suppose that one of the categories is empty. But concrete objects are indisputably real and well-understood, in contrast to abstract objects. So we should presume that abstract objects do not exist.[47]

This preference for shaving down one's ontology is directly related to the principle of Ockham's Razor—cutting away every assumption or hypothesis that can possibly be dispensed with. In this case,

---

46 Gould, "The Problem of God and Abstract Objects: A Prolegomenon," 271.

47 William Lane Craig, "Nominalism and Divine Aseity," in *Oxford Studies in Philosophy of Religion* 4, ed. Jonathan L. Kvanvig (New York: Oxford University Press, 2012): 48.

it is the assumption of the existence of abstract objects—or divine thoughts doing the work of abstract objects—that is trimmed off.

The nominalist shares the Platonist's conviction that necessarily existing things, like mathematical objects and their relations, cannot be said to have been created. Like Peter van Inwagen, some contemporary Christian Platonists bite the bullet and accept the existence of uncreated entities that exist necessarily and eternally "alongside" God and then attempt—in one way or another—to reconcile this fact with God's aseity. The nominalist, on the other hand, deals with this challenge by denying the existence of such things wholesale. Simply put, the fact that necessarily existing entities cannot properly be said to have been created is part of their motivation for embracing nominalism. Craig explains, "It is not the existence of abstract objects as such, or even of free abstract objects, that poses a serious challenge to divine aseity but rather the putative existence of uncreatables."[48] In his view, creatability is what is needed to make sense of abstract objects having ontological dependence upon God.

What would nominalism mean for God having—as Kepler believed—a preconceived, mathematically describable plan for creation? Craig regards the two ideas as entirely compatible:

> It seems to me that, whether one is a realist or an anti-realist about mathematical objects, the theist enjoys a considerable advantage over the naturalist in explaining the uncanny success of mathematics. On the one hand, God has created the world according to a certain blueprint which He had in mind. He might have chosen any number of blueprints. The world exhibits the mathematical structure it does because God has chosen to create it according to the model He had in mind...Thus, the theist—whether he be a realist or an anti-realist about mathematical objects—has the explanatory resources to account for the mathematical structure of the physical world. On the other hand, the theistic anti-realist also

---

48 Craig, "Nominalism and Divine Aseity,." 48.

has a ready explanation…God has created the world according to a certain blueprint which He had in mind.[49]

Elsewhere, in a discussion about a species of nominalism known as fictionalism, Craig and Paul Copan explain that according to the Platonist account, "God has fashioned the world on the model of certain mathematical forms, [thus] mathematics is inherent in any accurate description of the world," but then point out that "a similar account is available to the fictionalist. God may have employed certain mathematical fictions to serve as a blueprint for his construction of the physical universe, and hence those fictions are useful as a descriptive framework for our empirical science."[50] However, it is unclear what it would mean, exactly, for God to employ a fiction in His mathematically describable plan for the sensible world and how God having mathematical ideas in mind does not reduce to theistic conceptualism with regard to such ideas.

Craig's motivation for embracing nominalism seems to be what he sees as shortcomings of the realist alternatives, and much of his defense of nominalism itself involves a complex critique of the neo-Quinean Indispensability Argument (as he articulates it). However, the aforementioned problems Craig sees with realist alternatives to Platonism do not seem insuperable, and those who affirm theistic Platonism or any of the realist alternatives may well have other reasons for their realism besides any perceived merits of the neo-Quinean argument.[51]

---

49   Craig, *God Over All*, 164.

50   Paul Copan and William Lane Craig, *Creation Out of Nothing: A Biblical, Philosophical, and Scientific Exploration* (Grand Rapids: Baker Academic, 2004), 183. Here, by "mathematical fictions" Craig seems to mean nominalistic particulars.

51   For instance, Yandell, in his essay defending Platonism in Gould's volume, does not mention the neo-Quinean argument. Also, Paul Gould, in his PhD dissertation at Purdue University, "A Defense of Platonic Theism" (2010), does not rely upon the Quinean argument to make his case. More recently, he has affirmed that his main motivation for Platonism is the Problem of Universals.

The main difficulty for nominalism, it seems, is how to account for the necessary truth of mathematical propositions such as 7 > 5. After all, was this not necessarily true before there were human beings to construct the statement, "Seven is greater than five"? Craig's response to this objection is as follows:

> It seems to me that what the anti-platonist should say is that during the Jurassic Period or in worlds in which God alone exists, it is not, strictly speaking, true that 7 > 5. Rather what we should affirm is simply that during the Jurassic Period, 7 > 5, and that in all possible worlds 7 > 5. This is in line with a deflationary theory of truth according to which the truth-predicate "is true" is simply a device of semantic ascent which enables us to talk about a statement rather than to assert the statement itself.[52]

Yandell has a twofold response to this tactic. First, he points out that making sentences the bearers of truth values is problematic:

> It was true that there were dinosaurs before there were language speakers. This does not merely mean that had someone said "There are dinosaurs" when there were, what they said would have been true. That itself requires that there were in fact dinosaurs then. On Craig's suggestion, it was true that dinosaurs existed only after the dinosaurs were [extinct].[53] Incidentally, this runs smack against the deflationist view that "There are dinosaurs" is true if and only if there are dinosaurs—not if and only if there is someone to say there are.[54]

This critique seems right. Would Craig really go as far as saying that the proposition "God exists as one Godhead in three persons" was not true, strictly speaking, before humans came on the scene

---

52 Craig, "Response to Keith Yandell," in Gould, *Beyond the Control of God?*, 41.

53 The actual word used in the text is "distinct," but the intended term is clearly "extinct."

54 Yandell, "Response to Critics: William Lane Craig," in Gould, *Beyond the Control of God?*, 48.

and thought up the proposition? Also consider the fact that before anyone made the statement "there are exactly five members of the set of Platonic polyhedra," it was true that a craftsman would not have been able to build a sixth one. In other words, there was a demonstrable truth that had simply not been thought or spoken. The other problem with Craig's response, according to Yandell, is that "one cannot deflate a modal proposition."[55] Taking a cue from Yandell's example: Necessarily, seven is greater than five if and only if seven is greater than five, *and things cannot be otherwise*. Says Yandell, "Without what follows 'and,' we leave out an essential element. The 'necessary' is not redundant, and it is short for 'necessarily true.'"[56] Thus, it does not seem that necessary abstract truths can be reconciled with a nominalist perspective without serious problems.

Kepler's View

Of all the discussed options for dealing with the problem of God and abstract objects, theistic conceptualism seems like a promising option from both a philosophical and theological perspective. It faces some issues that need further work, but they seem minor relative to the difficulties faced by nominalism. In addition, any alleged difficulties with theistic conceptualism's treatment of properties and propositions do not necessarily apply to mathematical objects and relations.[57] Although Kepler was by no means a metaphysician, perhaps he said enough about the mathematical archetypes for creation existing eternally within God's

---

55 Ibid.

56 Ibid.

57 Some philosophers, such as Thomas Morris and Christopher Menzel, construe numbers as a type of property. I do not agree; I follow Craig on seeing mathematical objects (including numbers) as distinct from properties. See Craig, *God Over All*, 3.

mind to classify his view, in contemporary terms, as theistic conceptualism, while recognizing that it is unknown what he would have said about possibilia, properties, and propositions. Moreover, Plantinga and Craig affirm that theistic conceptualism is the traditional Christian perspective; as previous chapters demonstrated, it was the one held by some prominent Church Fathers (notably Augustine), and the one that dominated medieval theology.[58] Recall Augustine's insistence that mathematical truths are grounded in the divine intellect and discovered by man, who is made in God's image. He asked, "who would dare to say that God has created all things without a rational plan?" and insisted that this plan "must be thought to exist nowhere but in the very mind of the Creator."[59] This seems quite consistent with Kepler's view, which makes perfect sense in light of his intellectual heritage.

Note that tentatively classifying Kepler's view as theistic conceptualism is not to definitively rule out its general compatibility with any other realist view. In fact, Kepler's understanding of an archetype existing from eternity prior to God's creation of the universe (the copy) harmonizes quite well with Gould's Biggest Bang and Bigger Bang ideas. However, anti-realism seems to be at odds (at least in some key respects) with Kepler's understanding of what it means for God to be the wellspring of eternal mathematical objects that are made manifest in the features of the material universe.

---

58   Craig, *God Over All*, 73.

59   Augustine, *Eighty-three Different Questions* in *The Fathers of the Church* 70, 80–81.

CHAPTER 11

Copy—The Mathematical Intelligibility of the Cosmos

The scientific enterprise critically depends upon nature having an orderly structure that is accessible to the human mind. Put another way, science is made possible by a special compatibility between the rational character of the cosmos, abstract mathematical tools, and the human mind. It is this mysterious situation that Kepler's tripartite harmony of archetype, copy, and image explains so well. This chapter will investigate the mathematical intelligibility of the cosmos—why numbers and their relationships can be used to devise mathematical theories that allow human investigators to unlock the secrets of nature.[1] As will be seen, great thinkers of the later Western Tradition (both theists and non-theists), some of whom were instrumental in the twentieth-century physics revolution, marveled at the intellectual transparency of nature—a situation that has been called "unreasonable," "mysterious," and even "miraculous," for it seems to be, in principle, outside the explanatory scope of science and irreconcilable with naturalism.

This chapter will begin with an evaluation of select writings by prominent figures of twentieth-century physics who wrote about the intelligibility of the cosmos: Max Planck, Albert Einstein, Ar-

---

1 Recall that Kepler understood the sensible world as "the corporeal image of God," constructed according to mathematical archetypes that are imprinted upon the human mind. This is the Keplerian explanation for why the universe is mathematically intelligible to man.

thur Eddington, and Eugene Wigner. Next, relevant aspects of the contemporary conversation will be considered, including the indispensability of mathematics in scientific discovery as well as the persistent philosophical questions surrounding cosmic intelligibility. Some scientists and philosophers of today openly regard this situation as uncanny, though they differ, of course, on the question of how to best explain it. The central aim of this chapter is to show that the second strand of Keplerian natural theology—the idea that the universe is the material manifestation (the copy) of the archetypal plan man was intended to discover—is a more intellectually satisfying explanation than naturalism can ever offer.

Philosophical Ponderings of Twentieth-Century Physicists

Max Planck (1858–1947), the German theoretical physicist who came to be known as the founding father of quantum theory, was awarded the Nobel Prize in 1918 for his discovery of energy quanta.[2] Planck postulated that electromagnetic energy exists in individual units he called quanta instead of continuous waves, and he devised a mathematical equation that could be used to represent the relationship between the energy of a quantum and frequency:

$$E = h\nu$$

E represents energy, ν (the Greek letter *nu*) is the frequency of the emitted waves measured in waves per second, and h is a constant—now known as Planck's constant.[3] Thus, if the frequency of the

---

2   Quantum theory deals with how matter and energy behave at the atomic and subatomic levels. Planck's discovery was made in 1900.

3   Planck worked out the numerical value of his constant (h) to a degree of accuracy that is close to the value employed in physics calculations today.

electromagnetic radiation is known, then its energy can be calculated. Planck's work facilitated the emergence of quantum mechanics, which radically transformed the field of physics.

Throughout the remainder of his career as a professor at the University of Berlin, Planck continued to make important contributions to theoretical physics, but in his later years he became increasingly interested in philosophical and religious questions. During the final decade of his life, he penned a series of essays that were collected and published posthumously under the title *Scientific Autobiography and Other Papers*. He opens the first essay, "Scientific Autobiography," with this revelation:

> My original decision to devote myself to science was a direct result of the discovery which has never ceased to fill me with enthusiasm since my early youth—the comprehension of the far from obvious fact that the laws of human reasoning coincide with the laws governing the sequences of the impressions we receive from the world about us; that, therefore, pure reasoning can enable man to gain an insight into the mechanism of the latter. In this connection, it is of paramount importance that the outside world is something independent from man, something absolute, and the quest for the laws which apply to this absolute appeared to me as the most sublime scientific pursuit in life.[4]

Planck recognized the compatibility between the mathematical, law-governed structure of nature and what he calls the "laws of human reasoning." He was captivated by the "far from obvious fact" that there should be such a fortunate resonance between the fundamental workings of the natural world and human rationality. This wondrous gift was the impetus for his commitment to a life of science.

---

4   Max Planck, *Scientific Autobiography and Other Papers* in Great Books of the Western World 56 (Chicago: Encyclopaedia Britannica, Inc., 1990), 77.

Later in the same essay, Planck discusses the precise mathematical measurements that are used in physics and the significance of the constants of nature (such as his *h*) which are discovered through investigation. He writes, "These minute numbers, the so-called universal constants, are in a sense the immutable building blocks of the edifice of theoretic physics."[5] He then asks a penetrating question: "What is the real meaning of these constants? Are they, in the last analysis, inventions of the inquiring mind of man, or do they possess a real meaning independent of human intelligence?"[6] Planck discards the positivist view that the mathematical constants of nature are merely useful fictions devised and employed by the physicist, and that since measurements presuppose an observer, the substance of any law of physics cannot be detached from the observer. Planck argues that the result of this view is that a law "loses its meaning as soon as one attempts mentally to eliminate the observer and to see something more, something real, behind him and his measurement."[7] The problem he sees with this idea is that it "disregards a circumstance which is of a decisive importance in the extension and progress of scientific knowledge": the reproducibility of physical measurements, which are independent of the individual experimenter and of when and where the measurements are taken.[8] For Planck, this is an indication that "the factor which is decisive for the result of the measurement lies beyond the observer, and that one is therefore necessarily led to questions concerning real causal connections operating independently of the observer."[9] In other words, the consistency with which mathematical constants appear in the laws of nature says

---

5 Max Planck, *Scientific Autobiography and Other Papers*, 113.
6 Ibid.
7 Ibid.
8 Ibid.
9 Ibid.

something about the objective deep structure of material reality. He goes on:

> Of course, even today a consistent positivist could call the universal constants mere inventions which have proved to be uncommonly useful in making possible an accurate and complete description of the most diversified results of measurements. But hardly any real physicist would take such an assertion seriously. The universal constants were not invented for reasons of practical convenience, but have forced themselves upon us irresistibly because of the agreement between the results of all relevant measurements and—this is the essential thing—we know quite well in advance that all future measurements will lead to these selfsame constants. To sum it all up, we can say that physical science demands that we admit the existence of a real world independent from us, a world which we can however never recognize directly but can apprehend only through the medium of our sense experiences and of the measurements mediated by them.[10]

Clearly, Planck sees the constants of nature as legitimate discoveries about nature's mathematical structure rather than artificial, human-imposed ordering principles. As he so eloquently puts it, "they have forced themselves upon us irresistibly"; we can indeed glean true knowledge about the physical world, which seems to be inherently mathematical.

In a passage that has a Keplerian ring to it, Planck expresses his astonishment at the fact that man, tiny as he is in comparison with the whole wide universe, is able to at least partially comprehend it:

> In fact, how pitifully small, how powerless we human beings must appear to ourselves if we stop to think that the planet Earth on which we live our lives is just a minute, infinitesimal mote of dust; on the other hand how peculiar it must seem that we, tiny

---

10 Ibid., 114.

creatures on a tiny planet, are nevertheless capable of knowing though not the essence at least the existence and the dimensions of the basic building blocks of the entire great Cosmos![11]

Planck remarks that the natural laws the physicist uncovers give "the impression in every unbiased mind that nature is ruled by a rational, purposive will."[12] He articulates his expectation that the progress of the natural sciences will advance mankind's insights about "the omnipotent Reason which rules over Nature," and adds that "the deity which the religious person seeks to bring closer to himself by his palpable symbols, is consubstantial with the power acting in accordance with natural laws for which the sense data of the scientist provide a certain degree of evidence."[13] Planck's meaning seems to be that the rational laws of nature, which the scientist works to discern, are dictated by a Law Giver.

It is well worth noting that Planck explicitly denies any inherent conflict between science and religious belief; in fact, he argues that "both religion and natural science require a belief in God for their activities."[14] Planck's conviction is that there is a synergy between theism and the natural sciences, that they have the common goal of illuminating reality, and that the ultimate truth towards which these parallel streams are flowing is God. He writes:

> No matter where and how far we look, nowhere do we find a contradiction between religion and natural science. On the contrary, we find a complete concordance in the very points of decisive importance. Religion and natural science do not exclude each other, as many contemporaries of ours would believe or fear; they

---

11 Planck, *Scientific Autobiography and Other Papers*, 114. See discussion in Chapter 7. Kepler remarked that man is "puny" compared with the rest of the cosmos, but that he is nonetheless uniquely privileged in his ability to investigate nature's mathematical harmonies.

12 Ibid., 115.

13 Ibid., 116.

14 Ibid.

mutually supplement and condition each other. The most immediate proof of the compatibility of religion and natural science, even under the most thorough critical scrutiny, is the historic fact that the very greatest natural scientists of all times—men such as Kepler, Newton, Leibniz—were permeated by a most profound religious attitude.[15]

Planck describes the joint effort of science and religion as a crusade against both skepticism and superstition; he ends his essay with the battle cry, "On to God!"[16]

After serving as a university professor in both Zürich and Prague, Albert Einstein (1879–1955) accepted an invitation to join Max Planck and an elite group of scientists at the University of Berlin in 1914. Einstein is best known for his theories of special and general relativity, which transformed the scientific understanding of space, time, and gravity. He formulated the famous equation that expresses the mathematical relationship between mass and energy:

$$E = mc^2$$

where E represents energy, m is the mass of a quantity of matter, and c is the velocity of light. (One can only imagine how this mathematical interconnection—this harmony—of mass, energy, and velocity would have thrilled Kepler!) Einstein was awarded the 1921 Nobel Prize in Physics for his discovery of the photoelectric effect—the dislodging of an electron from a metal surface when a high-energy light photon (a quantum) collides with it. This was a major contribution to the emerging field of quantum mechanics, but (ironically) the mathematics involved conflicted with his theory of general relativity. Einstein never wavered from his conviction that the cosmos is a mathematically coherent whole, and his aspiration for the remainder of his life was to discover a grand unified

---

15   Ibid., 117.

16   Ibid.

theory—a goal he never achieved, and which remains the holy grail of theoretical physics.

Like Planck, Einstein had an acute fascination with the mathematical comprehensibility of nature and the related metaphysical implications. He perceived some sort of inscrutable divinity in the rational organization of the universe and was amazed that human beings can grasp even a small portion of it:

> We are in the position of a little child entering a huge library filled with books in many languages. The child knows someone must have written those books...The child simply suspects a mysterious order in the arrangement of the books but doesn't know what it is. That, it seems to me, is the attitude of even the most intelligent human being toward God. We see the universe marvelously arranged and obeying certain laws but only dimly understand these laws. Our limited minds grasp the mysterious force that moves the constellations.[17]

It is interesting that Einstein uses a library filled with books in various languages as a metaphor for the order of the cosmos—the order the physicist works to discern; it evokes the Church Fathers, the giants of the scientific revolution (including Kepler), and others who spoke of God's self-revelation in the book of nature. He more explicitly mentions the themes of a rationally structured universe and its correspondence with the human mind in a series of letters to Maurice Solovine, a Romanian philosopher and mathematician: "I have never found a better expression than 'religious' for this trust in the rational nature of reality and of its peculiar accessibility to the human mind."[18] He is careful to distinguish between the idea of the human mind *imposing* order on the cosmos and *discovering* order in it:

---

17   Denis Brian, *Einstein: A Life* (New York: John Wiley & Sons, 1996), 186.

18   Albert Einstein, *Letters to Maurice Solovine*, ed. Neil Berger (Paris: Gauthier-Villars, 1956), 102–103.

Well, *a priori* one should expect a chaotic world which cannot be grasped by the mind in any way. One could (yes *one should*) expect the world to be subjected to law only to the extent that we order it through our intelligence. Ordering of this kind would be like the alphabetical ordering of the words of a language. By contrast, the kind of order created by Newton's theory of gravitation, for instance, is wholly different. Even if the axioms of the theory are proposed by man, the success of such a project presupposes a high degree of ordering of the objective world, and this could not be expected *a priori*. That is the "miracle" which is being constantly reinforced as our knowledge expands.[19]

Here, Einstein rightly highlights the fact that there is simply no reason to expect the universe to be arranged in a manner that the human intellect can grasp and insists that even if human scientists devised the mathematical starting point from which to investigate, a high level of inherent orderliness of a peculiar kind is required for advanced mathematical physics to work.

Although he seems to reject all conceptions of God as a supreme being interested in human affairs, Einstein admits to having a kind of religiosity characterized by "a humble admiration of the infinitely superior spirit who reveals himself in the slight details we are able to perceive with our frail and feeble minds."[20] In several of his writings he speaks of his "rapturous amazement at the harmony of natural law, which reveals an intelligence of such superiority that, compared with it, all systematic thinking and acting of human beings is an utterly insignificant reflection."[21] Again, we see a direct connection to the theme of man's intelligence being somehow analogous to, and resonant with, the fundamental rationality of the universe. In a well-known 1936 essay entitled "Physics and Reali-

---

19 Ibid., 117.

20 Max Jammer, *Einstein and Religion* (Princeton, NJ: Princeton University Press, 1999), 93. For discussion on Einstein's understanding of the divine, see Endnote 18.

21 Albert Einstein, *The World as I See It* (New York: Kensington, 2006), 31.

ty" Einstein declares of the universe: "The fact that it is comprehensible is a miracle."[22] In the aforementioned letter to Solovine he writes: "There lies the weakness of positivists and professional atheists who are elated because they feel that they have not only successfully rid the world of gods but 'bared the miracles.' Oddly enough, we must be satisfied to acknowledge the 'miracle' without there being any legitimate way for us to approach it."[23] It is notable that Einstein seems to be admitting that this is a question outside the purview of scientific investigation. As will be seen, some contemporary thinkers would do well to recognize this limitation of the natural sciences.

Sir Arthur Stanley Eddington (1882–1944), who is regarded as the father of modern theoretical astrophysics, was an English contemporary of Planck and Einstein. Beyond providing eclipse photographs that served as empirical confirmation of Einstein's theory of general relativity, Eddington established himself as the founder of the discipline of stellar dynamics with the publication of his revolutionary work *Stellar Movements and the Structure of the Universe* (1914).[24] Later, he wrote what Einstein declared to be the best treatment of relativity in any language: *The Mathematical Theory of Relativity* (1923), the first explication of relativity theory in English. Eddington was fascinated with questions surrounding cosmic origins, and wrote a popular-level book, *The Expanding Universe*

---

22 Albert Einstein, "Physics and Reality," *Daedalus* 132, no. 4 (Fall, 2003): 24.

23 Einstein, *Letters to Maurice Solovine*, 119.

24 Eddington's photos of a 1919 solar eclipse showed that the sun's gravity bends beams of starlight when they pass by it in transit to the earth. This effect is known as parallax. Photos of a 1922 eclipse provided additional confirmation of Einstein's gravitational theory.

(1920), that explained and critiqued Lemaître's "fireworks theory" (later known as the Big Bang theory).[25]

The dramatic advancements being made in theoretical physics and astrophysics during Eddington's career highlighted the fact that mathematics functions as a tool for scientific discovery. Eddington was so impressed by the utility of mathematical equations in astrophysics that he often made remarks to that effect. In *The Expanding Universe*, he explains that by uniting the mathematical equations of quantum theory and those used in wave mechanics, astrophysicists can determine the rate at which spiral nebulae (what we now call spiral galaxies) are receding from the earth-bound observer's perspective: "By combining the two theories," he writes, "we can make the desired theoretical calculation of the speed of recession."[26] Eddington emphasizes that this process is entirely mathematical: "No astronomical observations of any kind are used in this calculation, all the data being found in the laboratory. Therefore when we turn our telescopes and spectroscopes on the distant nebulae and find them to be receding at a speed within these limits the confirmation is striking."[27] The stunning fact Eddington is emphasizing here is the predictive utility of mathematics in astrophysics; equations reveal truths about nature, sometimes even prior to the collection of observational data. Confidence in the precision with which these equations describe nature is enhanced by the fact that they are applicable to more than one question, and physical measurements can corroborate the purely mathematical results:

> If the theoretical ideas here employed had had only one application, viz. to calculate the recession of the nebulae, there might be

---

25 Arthur Eddington, *The Expanding Universe* (New York: Cambridge University Press, 1988), 56–57. Eddington suggests a theory that bears some resemblance to contemporary attempts at avoiding an ultimate beginning of all things (but he was not motivated by materialism). See Endnote 19.

26 Ibid., 93.

27 Ibid., 94.

a certain amount of room for "fudging." As a matter of fact the danger of unconscious fudging is greatly exaggerated; there is an artistry in these fundamental equations of physics which one cannot trifle with. But it naturally strengthens our confidence if the same step also leads to the solution of another problem. This happens in the present case, the associated problem being the relation of the proton to the electron and in particular the ratio of their masses.[28]

Eddington regarded this applicability of mathematics with wonder; that the mathematician sitting in the lab can predict, with accuracy, the recession rate of enormous galaxies and that the equations would also be useful for elucidating something about elementary particles was, for him, extraordinary. He was convinced that all the values of nature's constants that can be expressed in real numbers could be discovered mathematically, and his own calculations for constants such as the number of particles in the universe, the speed of light, the recession velocity of galaxies, and the ratio of gravitational force to electrical force between a proton and an electron were published posthumously in his *Fundamental Theory* (1946).[29]

Eddington was born into a Quaker family and remained in that faith tradition his entire life. This affiliation categorized him as a "Dissenter" among the Anglican-dominated intellectual elite of England.[30] He was, as historian of science Matthew Stanley puts it, "a world-class scientist who not just maintained his religious beliefs, but who also brought together the religious and scientific aspects of his life in powerful, meaningful, and productive ways."[31] However, Eddington did not exactly look favorably upon attempts

---

28   Eddington, *The Expanding Universe*, 95.

29   Eddington's work on these calculations anticipated later discussions on cosmic fine-tuning.

30   Matthew Stanley, *Practical Mystic: Religion, Science, and A.S. Eddington* (Chicago: University of Chicago Press, 2007), 7

31   Ibid., 2.

to prove the existence of God using scientific findings. In *The Nature of the Physical World* (1928) he writes:

> I repudiate the idea of proving the distinctive beliefs of religion either from the data of physical science or by the methods of physical science…I have sometimes been asked whether science cannot now furnish an argument which ought to convince any reasonable atheist. I could no more ram religious conviction into an atheist than I could ram a joke into the Scotchman.[32] The only hope of "converting" the latter is that through contact with merry-minded companions he may begin to realise that he is missing something in life which is worth attaining. Probably in the recesses of his solemn mind there exists inhibited the seed of humour, awaiting an awakening by such an impulse. The same advice would seem to apply to the propagation of religion…We cannot pretend to offer proofs.[33]

Eddington's rejection of any attempt to prove tenets of religion with science was partly due to his recognition of the fact that the conclusions of science change over time: "The lack of finality of scientific theories would be a very serious limitation of our argument, if we had staked much on their permanence. The religious reader may well be content that I have not offered him a God revealed by the quantum theory, and therefore liable to be swept away in the next scientific revolution."[34] Eddington did, however, acknowledge metaphysical implications of the fact that nature has a

---

32   Here, he refers to the "proverbial Scotchman with strong leanings towards philosophy and incapable of seeing a joke." See Eddington, *The Nature of the Physical World*, (New York: Cambridge University Press, 1958) 335–336.

33   Eddington, *The Nature of the Physical World*, 333, 336–337. This sentiment strikes one as quite Chestertonian; G. K. Chesterton definitely read Eddington, and once remarked, "It was Eddington I think, who used the phrase that the universe seems to be more like a great thought than a great machine." See G. K. Chesterton, *The Well and the Shallows* in *The Collected Works of G. K. Chesterton*, ed. James J. Thompson (San Francisco: Ignatius Press, 1990), 395.

34   Eddington, *The Nature of the Physical World*, 353.

mathematical order: "The harmony and simplicity of scientific law appeals strongly to our aesthetic feeling. It illustrates one kind of perfection, such as we might perhaps think worthy to be associated with the mind of God."[35] In *The Nature of the Physical World* he says, "The idea of a universal Mind or Logos would be, I think, a fairly plausible inference from the present state of scientific theory; at least it is in harmony with it."[36] He then cautions that all that may be suggested by nature's mathematical structure is "a purely colourless pantheism" devoid of any indication of the goodness or evilness of the mind behind the cosmos.[37] Nevertheless, he insists that "the stuff of the world is mind-stuff."[38]

Eugene Wigner and the "Unreasonable Effectiveness" of Mathematics

Hungarian physicist Eugene Wigner (1902–1995) was keenly interested in the work of Planck, Einstein, and others involved with the infant field of quantum theory, and he invested six decades of his life in the study of crystals, molecules, atoms, and atomic nuclei. A highly celebrated scientist in his own right, he won the Max Planck Medal in 1961 and the National Medal of Science in 1969. He was awarded the Nobel Prize in Physics in 1963 "for his contributions to the theory of the atomic nucleus and the discovery and application of fundamental symmetry principles."[39] He contribut-

---

35 Arthur Eddington, *Science and the Unseen World* (New York: Macmillan Company, 1929), 51.

36 Eddington, *The Nature of the Physical World*, 338.

37 Ibid.

38 Ibid., 276.

39 Wigner was awarded one half of the prize. The rest was divided equally between Maria Goeppert Mayer and J. Hans D. Jensen.

ed to the progress of relativity theory and quantum mechanics, and is also known for his essential work in designing nuclear reactors for the US government during World War II. Wigner came from a nominally Jewish family that, while he was still a boy, converted to Lutheranism for political reasons. He never developed much of a personal faith, lacking conviction one way or the other about the existence of God, but he maintained a gracious attitude toward the faithful and did not entirely avoid church attendance. In the autobiography he dictated near the end of his life he said, "Today, I am only mildly religious. When I attend church, it is with the Protestants."[40] Throughout his career, he found common science-related arguments for the existence of God uncompelling; he reports that in response to those claiming to have evidence showing that God made the world, he would sometimes quip, "Well, how...? With an earth-making machine?"[41]

Wigner's lack of religious conviction is notable for present purposes because of how central one of his essays has been in subsequent conversations about the theistic implications (or lack thereof) of the mathematical orderliness of nature. In 1960, he published "The Unreasonable Effectiveness of Mathematics in the Natural World," a philosophical exposition on the mysterious fact that mathematics is crucial to the investigation of the cosmos. In the essay, Wigner discusses the astounding degree to which mathematics maps on to the material world, pointing out that "mathematical concepts turn up in entirely unexpected connections. Moreover, they often permit an unexpectedly close and accurate description of the phenomena in these connections."[42] He insists that "the enor-

---

[40] Eugene Wigner and Andrew Szanton, *The Recollections of Eugene P. Wigner* (Cambridge, MA: Basic Books, 2003), 39.

[41] Ibid., 60.

[42] Eugene Wigner, "The Unreasonable Effectiveness of Mathematics in the Natural Sciences," reprinted in *The World Treasury of Physics, Astronomy, and Mathematics*, ed. Timothy Ferris (Boston: Little, Brown & Co., 1991), 527.

mous usefulness of mathematics in the natural sciences is something bordering on the mysterious…there is no rational explanation for it."[43] When the physicist formulates mathematics to characterize his laboratory observations, Wigner explains, it frequently leads to a surprisingly accurate description of a broader class of physical phenomena. "This shows," he says, "that the mathematical language has more to commend it than being the only language which we can speak; it shows that it is, in a very real sense, the correct language."[44] In other words, it is not a case of human scientists devising merely one effective tool; the level of precision involved strongly indicates that our mathematics is the *necessary* tool.

Wigner was especially intrigued by the fact that sometimes complex mathematical systems that were developed independently of any thought to physical applicability later turn out to be exceptionally useful in the physical sciences much like a lock-and-key fit. To use an analogy, this situation is like manufacturing an intricately shaped key sheerly for aesthetic purposes and then, many years later, coming across a lock that the old key is exclusively successful in opening. Although it is true that simple forms of mathematics were devised to measure aspects of the physical world, this is not always how things have happened; Wigner says that "whereas it is unquestionably true that the concepts of elementary mathematics and particularly elementary geometry were formulated to describe entities which are directly suggested by the actual world, the same does not seem to be true of the more advanced concepts, in particular the concepts which play such an important role in physics."[45] Pre-existing mathematical concepts that have been pressed into service to physics were previously created by the mathematician to "demonstrate his ingenuity and sense of formal beauty" without

---

[43] Eugene Wigner, "The Unreasonable Effectiveness of Mathematics in the Natural Sciences," 527.

[44] Ibid., 534.

[45] Ibid., 528.

any thought to potential scientific applicability.[46] In other words, the scientist first encounters a lock and *then* finds out that just the right key was constructed in the field of pure mathematics before any knowledge of the lock's existence.

Morris Kline, an American mathematician who was Wigner's contemporary, was also interested in the mathematics-nature connection. He offered a classic example of mathematics preceding scientific applicability in his work, *Mathematics and the Physical World* (published, *nota bene*, the year before Wigner's essay). Kline explains that the Greek mathematicians' explication of the curves of conic sections was pressed into service to physics well over one thousand years later when Kepler used the ellipse to describe planetary motion and Galileo used parabolas in his analysis of terrestrial projectile motion.[47] Alfred North Whitehead (1861–1947) had previously made the same observation in his work, *An Introduction to Mathematics*:

> ...conic sections were studied for eighteen hundred years merely as an abstract science, without a thought of any utility other than to satisfy the craving for knowledge on the part of mathematicians...then at the end of this long period of abstract study, they were found to be the necessary key with which to attain the knowledge of one of the most important laws of nature.[48]

The law of nature to which Whitehead refers is undoubtedly Kepler's First Law, which says that planets revolve around the sun in an elliptical path. Wigner notes that the mathematics of the ellipse went on to be key in Newton's formulation of the universal law

---

46   Ibid.

47   Morris Kline, *Mathematics and the Physical World* (New York: Dover Publications, 1959), 472. Apollonius of Perga (c. 260–200 BC) wrote a treatise entitled *On Conic Sections*, but the originator of conic section analysis was Menaechmus (c. 375–325 BC), a pupil of Plato. See Alfred North Whitehead, *An Introduction to Mathematics*, Great Books of the Western World 56 (Chicago: Encyclopaedia Britannica, Inc., 1990), 156.

48   Alfred North Whitehead, *An Introduction to Mathematics*, 158.

of gravitation, a law "formulated in terms which appear simple to the mathematician, which has proved accurate beyond all reasonable expectation."[49] Kline explains that developments in pure mathematics both anticipate the needs of science and even suggest solutions; he argues that it is "unlikely that Kepler would have invented the ellipse to describe planetary motion, because that task, together with the one he actually performed of fitting the ellipse to data, would have been superhuman."[50] He adds that "Einstein took full advantage of the already existing non-Euclidean geometry to create the theory of relativity," whereas the invention of an ingenious new geometry "and applying it to a radically new physical theory would have been beyond the powers of one man."[51] In both of these cases, the complex mathematical systems necessary for making major strides in physics were developed *independently*, without regard for their potential applicability to nature. Contemporary atheist physicist Steven Weinberg even admits, "It is positively spooky how the physicist finds the mathematician has been there before him or her."[52] In his book, *Dreams of a Final Theory*, Weinberg writes, "It is very strange that mathematicians are led by their sense of mathematical beauty to develop formal structures that physicists only later find useful, even where the mathematician had no such goal in mind...Physicists generally find the ability of mathematicians to anticipate the mathematics needed in the theories of physicists quite uncanny."[53]

---

49 Eugene Wigner, "The Unreasonable Effectiveness," 535.

50 Morris Kline, *Mathematics and the Physical World*, 473.

51 Ibid. Here, Kline refers to Riemannian geometry, which was developed decades before Einstein harnessed it for relativity theory.

52 Steven Weinberg, "Lecture on the Applicability of Mathematics," *Notices of the American Mathematical Society* 33.5 (Oct), quoted in Mark Steiner, *The Applicability of Mathematics as a Philosophical Problem* (Cambridge: Harvard University Press, 1998).

53 Steven Weinberg, *Dreams of a Final Theory: The Scientist's Search for the Ultimate Laws of Nature* (New York: Vintage Books, 1994), 157.

In his famous essay, Wigner briefly mentions what he considers another mystery directly related to the applicability of mathematics to nature: why the human mind is equipped to carry out the convoluted mathematical reasoning required in sciences such as quantum theory. He says that "certainly it is hard to believe that our reasoning power was brought, by Darwin's process of natural selection, to the perfection which it seems to possess."[54] In a particularly notable statement, Wigner remarks that "it is not at all natural that 'laws of nature' exist, much less that man is able to discover them."[55] He perceives the incredible compatibility between the higher cognitive capabilities of man and the rational mechanics of nature, which can be expressed in mathematical language.[56] He adds that these laws must be already formulated in mathematical language in order to be of use in the applied mathematics of the physicist, and that the point about nature's laws being written in such a language was "properly made three hundred years ago"—a reference to either Galileo or Kepler. Wigner muses: "It is difficult to avoid the impression that a miracle confronts us here, quite comparable in its striking nature to the miracle that the human mind can string a thousand arguments together without getting itself into contradictions or to the two miracles of the existence of the laws of nature and of the human mind's capacity to divine them."[57] He does not presume to have an explanation for this miracle; he seems resigned to regard it as permanently inscrutable:

> The miracle of the appropriateness of the language of mathematics for the formulation of the laws of physics is a wonderful gift which we neither understand nor deserve. We should be grateful for it and hope that it will remain valid in future research

---

54 Wigner, "The Unreasonable Effectiveness," 528.
55 Ibid., 531.
56 Ibid., 532.
57 Ibid., 533.

and that it will extend, for better or for worse, to our pleasure even though perhaps also to our bafflement, to wide branches of learning.[58]

Wigner stopped well short of suggesting divine design of the cosmos as an explanation, but it is not difficult to see why many have drawn a theistic conclusion from the "unreasonable effectiveness" phenomenon his essay explores. One may reasonably infer that there is a connection of some sort between mathematical truths and the material world, which seems to suggest that nature is in some way informed by a rational mind.[59] The consequence is that scientific investigation is possible, and this in turn seems to imply an anthropocentric world.

Wigner's perspective has not gone unchallenged. In a 2008 paper, for example, mathematical historian and logician Ivor Grattan-Guinness claims to have solved Wigner's mystery; he says that the applicability of mathematics is in fact reasonable, given the way mathematical theories have developed. He concludes that "*it is the world of human theories that is anthropocentric, not the actual world.*"[60] Contra Wigner, Grattan-Guinness contends that even the more advanced theories—such as those used in physics—have been "motivated by some problems found in the actual world, including on occasion sciences outside the physical ones."[61] "Much mathematics," he says, "at all levels, was brought into being by worldly demands, so that its frequent effectiveness there is not so surpris-

---

58 Wigner, "The Unreasonable Effectiveness," 528.

59 It is interesting that Wigner's brother-in-law, the famous theoretical physicist and avowed atheist Paul Dirac once remarked, "God is a mathematician of a very high order and He used advanced mathematics in constructing the universe." See Paul Dirac, "The Evolution of the Physicist's Picture of Nature," *Scientific American* 208, no. 5 (May 1963), 53.

60 Ivor Grattan-Guinness, "Solving Wigner's Mystery: The Reasonable (Though Perhaps Limited) Effectiveness of Mathematics in the Natural Sciences," *The Mathematical Intelligencer* 30, no. 3 (2008): 9. Emphasis his.

61 Ibid., 8.

ing."[62] But as mathematician Russell Howell has pointed out, "no physical phenomena guided the formation of complex analysis—a key tool for Wigner."[63] As previously discussed, there are important instances in which mathematics that turned out to be essential to the natural sciences was devised *independently* of empirical considerations. One might also respond to Grattan-Guinness's argument by pointing out that *even if* all mathematics used in the natural sciences was humanly devised for each particular purpose, such a situation is open to an alternative interpretation: that what really occurs in this human process is the extraction of mathematical concepts *from* the features of the material world. After all, how else could such effective mathematics be invented if they did not reflect a very real mathematical order in the structure of nature, as Einstein argued?

Another critique that Grattan-Guinness offers is the fact that mathematical theories are derived from or influenced by pre-existing theories: "Wigner...underrated the central place of theories being formed in the presence of other theories, and being desimplified when necessary and where possible...theory-building can be seen as reasonable to a large extent."[64] Grattan-Guinness believes that Wigner's situation in the early days of quantum theory might have been a case of lucky guesswork in terms of finding appropriate mathematics for the models: "Perhaps [quantum theory's] first practitioners struck lucky in analogising from the experiential celestial heavens to the highly nonexperiential atom."[65] Howell counters, "If guesswork is involved in science, it is interesting that, as a grand strategy, the bullseye so often is hit when the method employed rests on mathematical theories that invariably

---

62 Ibid.

63 Russell W. Howell, "The Matter of Mathematics," *Perspectives on Science and Christian Faith* 67 no. 2 (June 2015): 82.

64 Grattan-Guinness, "Solving Wigner's Mystery," 15.

65 Ibid.

grew out of human aesthetic criteria."⁶⁶ An excellent example is non-Euclidean Riemannian geometry, which was developed in an effort to perfect existing geometry yet turned out to be precisely what Einstein needed to elucidate the contours of space-time.

Even if rebuttals like those of Grattan-Guinness carry some weight (and even Howell believes that they do), the fact remains that higher mathematics, whatever its origin and ontology (i.e., Platonic or nominalistic), is tremendously well-suited as a language for the natural sciences—a language humans are cognitively equipped to utilize—and this cries out for explanation. As Kline explains:

> The major, inescapable fact, and the one that still has inestimable importance, is that mathematics is the method par excellence by which to investigate, discover, and represent physical phenomena. In some branches of physics it is...the essence of our knowledge of the physical world. If mathematical structures are not in themselves the reality of the physical world, they are the only key we possess to that reality...And the accuracy with which mathematics can represent and predict natural occurrences has increased remarkably since Newton's day.⁶⁷

As Einstein asked, how could something that is the product of human thought align so precisely with physical reality?

Mathematics, Kline explains, has furnished the great scientists of the scientific revolution and the physics explosion of the early twentieth century with marvelously applicable and indispensable tools.⁶⁸ "All of these highly successful developments," he says, "rest on mathematical ideas and mathematical reasoning. The question becomes inescapable. Why does mathematics work?"⁶⁹

---

66 Howell, "The Matter of Mathematics," 82. Here, Howell is summarizing (and agreeing with) an argument made by philosopher Mark Steiner.

67 Kline, *Mathematics and the Physical World*, 465.

68 Ibid.

69 Ibid.

Even if the concepts and axioms of mathematics are derived from observations of the physical world, there comes a point at which the mathematician takes over and uses pure mathematical reasoning to glean deeper insights about nature. "It may indeed be true that the deduced facts are necessary consequences of the axioms, and the latter derive from the physical world," writes Kline, "but that reasoning of a highly intricate type should produce physically serviceable knowledge is the mystery which demands resolution. Why should the physical world conform to the pattern of man's reasoning?"[70] Indeed, why is it that the higher reasoning powers of man can engineer a system from basic concepts and axioms, a system that turns out to shed light on the fundamental structure of nature? In a 1959 passage that sounds so similar to Wigner that one cannot help but wonder if Wigner's essay was inspired by Kline, Kline writes:

> It may be that the effectiveness of the mathematical representation and analysis of the physical world is as unexplainable as the very existence of the world itself and of man. It is nonetheless a phenomenon that we can accept and utilize to great advantage. Having found that this gift from an unknown donor can be so profitably employed, our civilization now seems set to exploit it to the hilt. If the attempt to understand why it works leaves us with an enigma, as does the smile of the Mona Lisa, this merely means that we have an intriguing subject for further study and contemplation.[71]

An interesting account of mathematics serving as a marvelously predictive tool in the natural sciences begins in the 1960s, shortly after the publication of Kline's book and Wigner's essay. Physicist Peter Higgs and several colleagues were working on the problem of how subatomic particles (specifically, the quarks that make up pro-

---

70 Ibid., 470.
71 Ibid., 472.

tons and neutrons) get their mass. At the time, the standard model suggested massless particles; assigning masses to them threw the mathematics of the model into chaos. For Higgs, this meant that mass must be an emergent property that results from some sort of particle interactions. He theorized that empty space is entirely permeated with a so-called scalar field with which particles interact and thereby acquire energy and thus mass (recall Einstein's famous equation that demonstrated the equivalence of mass and energy). This theoretical entity came to be known as the "Higgs field" and, despite its lack of empirical support, was gradually accepted by the physics community as a viable explanation; it allowed physicists to keep their aesthetically satisfying, consistent system of mathematical equations while accounting for particle mass.

Half a century later, in 2012, CERN (the European Organization for Nuclear Research in Geneva) was finally able to empirically test the Higgs field theory using the multi-billion-dollar Large Hadron Collider (LHC), a high-tech underground particle accelerator outfitted with a system of particle detectors. Calculations from Higgs field theory indicated that particle collisions inside of the LHC should result in reverberations that cause the hypothesized field to fling off a signature particle—the Higgs boson. On the Fourth of July, physicists at the LHC announced that the Higgs boson had been detected, and this piece of the standard model was confirmed. Essentially, Higgs's purely mathematical theory had accurately predicted the existence of an unobserved physical entity and its properties. (Max Tegmark of MIT has often remarked that the Higgs field was predicted with a pencil.)[72] Decades after their mathematical prediction, Peter Higgs and Francois Englert shared the 2013 Nobel Prize for their discovery.

Another fascinating example of mathematics predicting natural phenomena came much earlier, in 1916, when Karl Schwarzschild

---

72 Tegmark often makes a comment along these lines in lectures based upon his book, *Our Mathematical Universe*, such as the one given at the Royal Institution on January 30, 2014.

discovered solutions to Einstein's field equations that suggested a very strange configuration of matter. Schwarzschild discovered these solutions while studying Einstein's gravitational theory and doing his own work on calculations of artillery trajectories during his time at the Russian front during World War I. He sent his results to Einstein, who presented them to the Prussian Academy on Schwarzschild's behalf. The implication of the solutions was that if a star's mass is concentrated in a spherical region that is so small that dividing its mass by its radius exceeds a critical value, then space-time will warp so radically that anything (including light) that gets within a minimum distance of the star will not be able to escape its gravitational pull. Schwarzschild dismissed the implications of the mathematics, but after half a century of subsequent scientific work that supported the existence of this phenomenon, John Archibald Wheeler gave it a name—the *black hole*.[73] The mathematics had been correct in its prediction of these strange entities before empirical evidence became available.

---

73 For a precise history of this term, see Endnote 20.

Chapter 12

The Scientific Applicability of Mathematics in Contemporary Thought

Deep philosophical questions concerning the indispensability of mathematics in scientific discovery persist in the contemporary conversation. Some scientists and philosophers (both theists and non-theists) openly regard the applicability of mathematics to the natural sciences as remarkable, though they differ, of course, on the question of how to best explain it. The central aim of this chapter is to show that the second strand of Keplerian natural theology—the idea that the universe is the material manifestation (the copy) of the archetypal plan man was intended to discover—is a more intellectually satisfying explanation than naturalism is able to offer.

An Anthropocentric Plan

In his 1998 work, *The Applicability of Mathematics as a Philosophical Problem*, Mark Steiner argues that the use of mathematics in the hard sciences is inherently anthropocentric; he says that "relying on mathematics in guessing the laws of nature is relying on human standards of beauty and convenience," yet this approach has undeniably been used by physicists with great success.[1] The universe is thus remarkably comprehensible for the human scientist equipped with the intellectual capacity for higher mathematical analysis and a preference for elegant mathematical systems (such as the internally coherent system preserved by the Higgs field theory). Steiner regards this anthropocentric utility of mathematics as a serious challenge for materialism; why should abstract operations carried on in the human mind have such descriptive and (especially) predictive power in the material realm if both are—as the materialist alleges—entirely subject to the laws of physics and chemistry? Also, why is it that mathematics plays a role in novel scientific discoveries? For these reasons, he sees the success of the "grand strategy" of applying the entire structure of mathematical concepts (which is, he says, an anthropocentrically defined structure) to nature as the central issue to be explained: "The world really does look anthropocentric—in the limited sense that it is intellectually accessible to human research."[2]

Steiner contends that in order to refute his argument, "one must find a natural, or material, property of mathematics as such, and then show how this property accounts for the success of the mathematical discoveries" related to the natural sciences.[3] He makes the key point that a Darwinian account of how mathematics emerged in human prehistory is insufficient, and may actually end up confirming his argument. A preference for patterns in nature *might* have been preserved through natural selection, and such a preference *might* have resulted in favoring the same sorts of patterns in mathematical notation, but this says nothing about why mathematical notation is successful in the sciences.[4] Thus, attempts

---

[1] Mark Steiner, *The Applicability of Mathematics as a Philosophical Problem*, 7.

[2] Ibid., 9.

[3] Ibid., 8.

[4] Ibid., 8–9.

to solve the ontological question of mathematics using naturalistic evolutionary accounts do nothing to mitigate the problem of the applicability of mathematics to the physical world. Steiner rejects the Kantian solution which says that mathematics is simply a man-made lens through which we investigate the sensible world:

> Some philosophers prefer a Kantian account of mathematical discovery: the world is the way it is, in part because of our contribution to our own experiences. Mathematics is the lens through which we view the Universe, meaning the phenomenal, or experienced Universe (about things in themselves we know nothing). This is...a valid attempt to explain away the data, but it will have to come to grips with the nature of contemporary science, which deals with objects beyond the realm of spatiotemporal experience.[5]

Steiner's objection seems right; after all, how does a Kantian account make sense of the construction of incredibly advanced mathematical theories that explain and predict unobserved phenomena in particle physics, or of Einstein's successful application of Riemannian geometry to the invisible fabric of space-time? As Eugene Wigner pointed out, the advanced mathematics that is used in fields such as theoretical physics goes well beyond the application of more elementary mathematics to the sensible world. Thus, a Kantian explanation seems insufficient, to say the least. The so-called Wigner/Steiner puzzle—why mathematics is so precisely applicable to nature—seems to remain an intractable problem for naturalism, and even some naturalists have admitted as much.

---

5   Ibid., 9.

Penrose's Triangle

Contemporary physicist Roger Penrose (a non-theist) acknowledges the serious questions surrounding the reality of a mathematically comprehensible universe. In several books and essays, he has discussed naturalism's inability to explain the relationship between mathematics, the material world, and the human mind—the conjunction that makes the natural sciences possible. He calls this the "three worlds, three mysteries problem," and has developed a diagram to illustrate the situation.

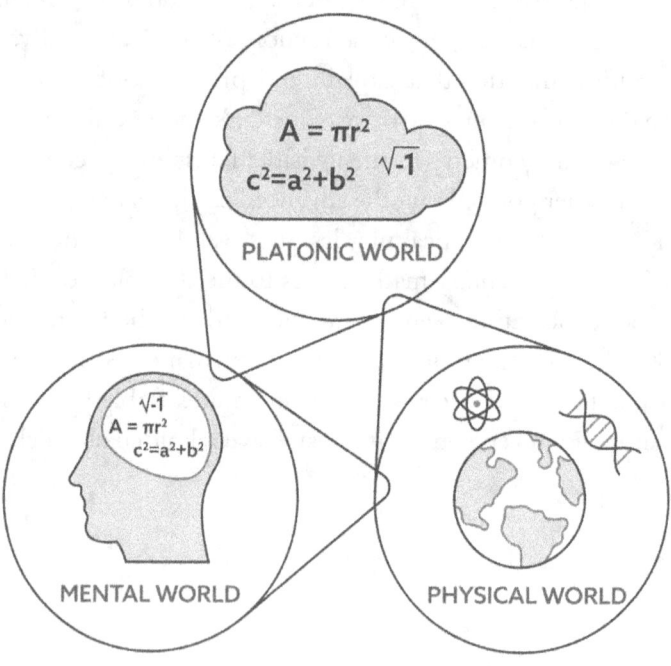

Figure 12.1 *Penrose's Triangle—Three Worlds, Three Mysteries*

Three spheres, each representing one of the three worlds—the Platonic world of mathematics, the material world, and the mental world—stand in triangular orientation to one another. The connecting cones, each one cross-sectioned near the apex, represents the small portion of each world that is involved with the next, the three "deeply mysterious connections" between these fundamentally different aspects of reality.[6] The diagram is meant to be read clockwise, beginning at the top. First, says Penrose, is the issue of the existence of mathematics, "whether the 'world of mathematics' arises merely as a product of our mental activities, having no reality beyond this, or whether it is to be assigned an independent, abstract existence of its own."[7] Penrose takes the latter view, believing that at least some degree of Platonism must be the case, since mathematical truths are eternal and objective.[8] The Platonic world, he explains, contains all mathematical entities such as the natural numbers, numerical relationships, and so on.[9] He writes that this realm

> may well seem to the reader to be just a rag-bag of abstract concepts that mathematicians have come up with from time to time. Yet its existence rests on the profound, timeless, and universal nature of these concepts, and on the fact that their laws are independent of those who discover them. The rag-bag—if indeed that is what it is—was not of our creation. The natural numbers were there before there were human beings, or indeed any other creature here on earth, and they will remain after all life has perished.[10]

---

6 Roger Penrose, "Mathematics, the Mind, and the Physical World," in *Meaning in Mathematics*, ed. John Polkinghorne (New York: Oxford University Press, 2011), 41.

7 Ibid.

8 Ibid., 44.

9 Roger Penrose, *Shadows of the Mind: A Search for the Missing Science of Consciousness* (New York: Oxford University Press, 1994), 412.

10 Ibid., 413.

Moreover, Penrose believes that the reality of the mathematical world is reinforced by Mystery 1, the fact that "the operations of the physical world are now known to be in accord with elegant mathematical theory to an enormous precision."[11] He gives the example of PSR 1913+16, the double-neutron-star system that was discovered in 1974, and explains that the agreement between the recorded timing of the system's pulsar signals and Einstein's general theory of relativity has a precision of approximately one hundredth of a thousandth of a second over a thirty-year period—an extraordinary concurrence.[12] Penrose regards Einstein's general theory of relativity as an impressive example of mathematical precision, one that seems to undercut some of the more popular counterarguments to Penrose's view:

> One not infrequently hears the viewpoint expressed that physicists are merely noticing patterns, from time to time, where mathematical concepts may happen to apply quite well to physical behaviour. It would be claimed, accordingly, that physicists tend to bias their interests in the directions of those areas where their mathematical descriptions work well, so there is no real mystery that mathematics is found to work in the descriptions that physicists use. It seems to me, however, that such a viewpoint is extraordinarily wide of its mark. It simply provides no explanation of the deep underlying unity that Einstein's theory, in particular, shows that there is between mathematics and the workings of the world...Einstein was not just 'noticing patterns' in the behavior of physical objects. He was uncovering a profound mathematical substructure that was already hidden in the very workings of the world. Moreover, he was not just searching around for whatever physical phenomenon might best fit a good theory. He found this

---

11   Penrose, "Mathematics, the Mind, and the Physical World," 44.
12   Ibid.

precise mathematical relationship in the very structure of space and time—the most fundamental of physical notions.[13]

Like Einstein, Penrose clearly rejects the idea that humans simply project mathematical order onto observable nature. Elsewhere he writes: "It makes no sense to me that this concurrence between the workings of the natural world at its most fundamental levels (here the very structure of space and time) and sophisticated mathematical theory is merely the result of our trying to fit the observational facts into some organizational scheme that we can comprehend."[14] He insists that "the concurrence between Nature and sophisticated beautiful mathematics is something that is 'out there' and has been so since times far earlier than the dawn of humanity, or of any other conscious entities that could have inhabited the universe as we know it."[15] In other words, mathematical truths and mathematical descriptions of natural phenomena are independent of human minds; they would hold even if no one ever thought of them. During the Cretaceous period, the Pythagorean theorem was true in the context of Euclidean geometry despite the absence of rational animals who could apprehend the theorem. For Penrose, the applicability of mathematics to nature strongly implies the objective reality of mathematics, but he realizes that mere Platonism is an insufficient explanation for Mystery 1.

Penrose's Mystery 2 is how the world of conscious perception could have emerged from the material world. He asks, "How does consciousness arise in a world which seems to be governed by entirely impersonal mathematical operations?"[16] In other words, when we observe the laws of nature and the cause-and-effect behavior of matter in motion, we must ask how it could have self-or-

---

13   Penrose, *Shadows of the Mind*, 415.

14   Penrose, "Mathematics, the Mind, and the Physical World," 45.

15   Ibid.

16   Ibid., 42.

ganized in such a way as to "wake up" and observe the world. Consciousness and rationality are completely different kinds of things that seem to transcend the material. Mystery 3 concerns how conscious minds are "mysteriously able (at least when they are at their best) to conjure up abstract mathematical forms, and thereby enable our minds to gain entry, by understanding, into the Platonic mathematical realm."[17] Both of these questions will be discussed extensively in the next chapter. For Penrose's part, he is amenable to the idea that the ultimate explanation of these mysteries may lie beyond the scope of modern physics; where the human mind is concerned, he submits that we may "have to look even farther afield, to an understanding that lies essentially beyond any kind of science whatever, as could be the implication of an essentially religious perspective."[18] Indeed, these are metaphysical conundrums; the natural sciences cannot even begin to unravel them.

Significantly, the deep connections between the three "worlds" of Penrose's triangle are essential to the natural sciences, which are made possible by the fact that the material world has a mathematically explicable structure that human minds can grasp. Why are things this way? For naturalists like Penrose, this seems to be an unsolvable mystery. However, some have attempted to resolve the triangle without abandoning naturalism or the independent existence of mathematical truths. Max Tegmark of MIT concurs with Penrose's Platonic realism; he is convinced that mathematics is an objective, mind-independent world that is simply "out there," and "any intelligent entity who begins to study any corner of it will inevitably discover at least the main plazas and connecting boulevards, even if many charming back alleys and sprawling suburbs are missed...[it] is thus the same whether it is discovered by us,

---

17 Penrose, *Shadows of the Mind*, 414.

18 Penrose, "Mathematics, the Mind, and the Physical World," 43.

by computers or by extraterrestrials."[19] However, to account for the connection between the mathematical and physical worlds that Penrose calls mysterious, Tegmark adopts what might be described as a thoroughly Pythagorean form of Platonism. He suggests that mathematical and physical existence are *equivalent*, that "the physical world really is completely mathematical, isomorphic to some mathematical structure."[20] Put another way, nature is not merely *described* by mathematics, it *is* mathematics, and human beings are actually "self-aware parts of a giant mathematical object."[21]

To support his thesis, which he calls the Mathematical Universe Hypothesis (MUH), Tegmark cites numerous examples of how mathematics pervades nature, sometimes to such an extent that it enables scientists to make correct predictions about material reality. He believes that if physicists eventually achieve a coherent Theory of Everything (ToE), its equations would be a comprehensive description of all reality. The structure of the ToE would be *equivalent* to the structure of external physical reality, leaving no excess "baggage" (properties immune to mathematical description) and thus demonstrating, according to Tegmark, that "our external physical reality and the mathematical structure are one and the same."[22]

The MUH is an excellent example of the extreme lengths some are willing to go to properly acknowledge the mathematical nature

---

[19] Piet Hut, Mark Alford, and Max Tegmark, "On Math, Matter and Mind," *Foundations of Physics* 36, no. 6 (June 2006): 769. I refer only to Tegmark because each author has contributed his own section of the paper.

[20] Ibid.

[21] Max Tegmark, *Our Mathematical Universe: My Quest for the Ultimate Nature of Reality* (New York: Random House, 2014), 6. The essential idea behind Tegmark's hypothesis is not original. Atheist and Oxford chemist Peter Atkins has also suggested that the universe is mathematics, a theory he calls *strong deep structuralism*. Tegmark does not credit Atkins. See Peter Atkins, *Creation Revisited* (London: Penguin Books, 1994), 109.

[22] Tegmark, *Our Mathematical Universe*, 280.

of the cosmos while maintaining their naturalistic worldview. Tegmark cites Galileo and Wigner as his intellectual predecessors:

> After Galileo promulgated the mathematical universe idea, additional mathematical regularities beyond his wildest dreams were uncovered, ranging from the motions of planets to the properties of atoms. After Wigner had written his 1967 [sic] essay, the standard model of particle physics revealed new "unreasonable" mathematical order in the microcosm of elementary particles and in the macrocosm of the early universe. I know of no other compelling explanation for this trend than that the physical world really is completely mathematical.[23]

At the end of this passage, Tegmark implicitly rules out the idea that the universe exhibits mathematical rationality because it is the product of a rational mind. In fact, he seeks to explain away mind itself as nothing more than mathematics. Yet, the MUH suffers from serious philosophical problems. First, it does not explain why this (hypothetical) grand mathematical object exists in the first place. Granted, this is a metaphysical question, but it is one that must be addressed. Second, how are we to make sense of mathematics having or instantiating physical existence? Philosopher Roger Trigg has described Tegmark's MUH as a scenario in which "mathematics calls the shots, becoming a rational force that mysteriously creates the physical from the purely conceptual and abstract," and points out that this is a metaphysical, not scientific, claim.[24] He continues his critique by citing what seems to be a fatal flaw of the MUH: "What counts as 'physical' is so far removed from any current understanding that it is indescribable in terms

---

23 Tegmark, "The Mathematical Universe," *Foundations of Physics* 38, no. 2 (February 2008): 108.

24 Roger Trigg, *Beyond Matter: Why Science Needs Metaphysics* (West Conshohocken, PA: Templeton Press, 2015), 46.

acceptable to current science."[25] Third and finally, the MUH's claim that the human mind—which is capable of mathematical reasoning and subjective experiences—is itself nothing more than complex mathematics in material form operating according to mathematical laws suffers from major philosophical difficulties that will be addressed in Chapter 13.

Philosophical Implications of the Comprehensible Universe

The meta-level question that falls outside of naturalism's explanatory scope is why the cosmos is intellectually accessible to human beings. Comprehensibility did not have to be the case; it is a contingent feature of the cosmos. Because mathematics is the tool *par excellence* for scientific investigation, some naturalists believe the explanation for mathematical comprehensibility lies entirely within the human brain, which is (they argue) a product of non-teleological evolutionary processes. George Lakoff and Rafael Núñez, for example, contend that mathematics exists only within the human mind as a result of the evolutionary development of the brain, and "whatever 'fit' there is between mathematics and the world occurs in the minds of scientists who have observed the world closely, learned the appropriate mathematics well (or invented it), and fit them together (often effectively) using their all-too-human minds and brains."[26] However, this ignores the higher question of why nature is rationally organized, and it is far from clear that the scenario Lakoff and Núñez offer is not merely a naturalistic interpretation of the situation. It is equally (if not more) plausible to understand

---

25   Ibid., 47.

26   George Lakoff and Rafael Núñez, *Where Mathematics Comes From* (New York: Basic Books, 2000), 3.

the scientific endeavor as human thinkers devising a coherent system of mathematical notation that enables them to grasp the rational structure of the universe. The *system of notation* they employ is certainly invented, but that is not to say that the *concepts and relations* the notations symbolize are not somehow inherent to the physical world. In other words, to have an account for how human beings have come to know the mathematics of nature says nothing about an objective rationality *in* nature; to suppose otherwise is to confuse *epistemology* (how something is known) with *ontology* (what objectively exists).

Astronomer and former director of the Vatican Observatory, George Coyne, and Michael Heller, a Templeton Prize-winning philosopher, regard the rationality of nature as a philosophical problem that does not submit to scientific analysis. They argue that intelligibility involves what they call an ontological rationality— "the conditions the world must satisfy to render science possible."[27] As Einstein remarked in his aforementioned famous essay, "Physics and Reality":

> The very fact that the totality of our sense experiences is such that by means of thinking (operations with concepts, and the creation and use of definite functional relations between them, and the coordination of sense experiences to these concepts) it can be put in order. This fact is one which leaves us in awe, but which we shall never understand. One may say "the eternal mystery of the world is its comprehensibility."[28]

Considering this old and enduring idea, Coyne and Heller have developed what they call the *hypothesis of the rationality of the world*. To arguments such as Lakoff and Núñez's, they respond that, in the

---

[27] George Coyne and Michael Heller, *A Comprehensible Universe: The Interplay of Science and Theology* (New York: Springer, 2010), 28.

[28] Albert Einstein, "Physics and Reality," 23–24.

sphere of physics, the artificial *imposition* of order is not a plausible explanation for the intrinsic rationality of nature:

> Mathematical models used in physics not only describe some aspects of the world, but they also imitate them in some sense; or more precisely, they model them, i.e. they function in a way similar to the investigated aspects of the world. In this sense mathematics is something more than just the language of physics. Some metaphorically say that it is "the stuff out of which physics is made"...you cannot construct a mathematical model of something that is irrational. Irrationality introduces into the model contradictory elements, and contradictions destroy the model.[29]

Put another way, if there was not a very real sense in which the cosmos has the property of rationality, mathematical models of its phenomena would break down. Yet, the precise opposite is the case; mathematics has proven to be an extraordinarily effective and unique catalyst for scientific discovery. Each time physics began applying mathematics on a new scale, which was the case with Kepler, Galileo, and Newton, and most assuredly the case again in early twentieth-century physics, the momentum of the natural sciences increased dramatically. Coyne and Heller write, "When physics started using [mathematics] on a large scale, the progress in understanding the world became so rapid that it cannot be paralleled by the progress in any other field of human activity"[30] They are convinced that this suggests a genuine connection between mathematics and nature, which means "we should ascribe to the world a property owing to which it can be efficiently investigated with the help of the mathematical-empirical method."[31] It is precisely this property that transcends scientific explanation and calls for philosophical and theological analysis.

---

29 Coyne and Heller, *A Comprehensible Universe*, 30.

30 Ibid., 30, 101.

31 Ibid., 30.

John Polkinghorne, a theoretical physicist[32] and member of the Royal Society who made a mid-life career shift to priesthood in the Anglican church (and science and religion scholarship), argues that, according to naturalism, the comprehensibility of the universe is a fortunate fluke that makes the natural sciences possible. He writes:

> Fundamental to the scientific enterprise is belief in the possibility of access to the basic structure of an intelligible universe. From a scientistic point of view, the possibility of deep science can only be regarded as resting on the incredibly happy accident that the universe has been found to have this fortunate character of rational transparency, and that we have proved to be persons capable of exploiting the opportunity that this affords. Strangest of all, it has turned out that the key to unlocking these deep secrets is provided by the abstract subject of mathematics.[33]

Naturalism simply does not have the explanatory resources to account for the intellectual accessibility of the material world because the issue is a metaphysical one. Those with a scientific bent, who are unwilling to explore the question on the appropriate meta-level, must be content with having no explanation at all. Polkinghorne goes on to say,

> Yet, what for the scientist, as a scientist, is simply the miraculous fact that the physical world is deeply and wonderfully intelligible, can become in its turn comprehensible in the light of theological understanding, since belief in God enables one to see cosmic

---

32 An interesting connection to Chapter 11—Polkinghorne earned his Ph.D. at Cambridge in a research group led by famous physicist Paul Dirac, Eugene Wigner's brother-in-law. See note 59 of Chapter 11.

33 John Polkinghorne, *Theology in the Context of Science* (New Haven: Yale University Press, 2009), 71–72.

intelligibility and the beautiful ordering of the universe as reflections of the rational character of the world's creator.[34]

According to theism, the precision with which mathematics illuminates physical reality makes perfect sense, because the world is conceived as the material manifestation (the Keplerian copy) of a rational plan (the Keplerian archetype) in the mind of God. It is a far more intellectually satisfying explanation than the highly improbable "happy accident." In this way, the argument from mathematical comprehensibility to theism is an inference to the best explanation. As philosopher Keith Ward has put it, "The continuing conformity of physical particles to precise mathematical relationships is something that is much more likely to exist if there is an ordering cosmic mathematician who sets up the correlation in the requisite way."[35]

Thus, the "unreasonable effectiveness" of mathematics to the natural sciences has strong theistic implications. However, the issue goes even deeper; the mathematical comprehension of nature is also contingent upon the existence of rational creatures who are able to discern nature's order through the construction and application of advanced mathematical systems. This involves the third and final strand of Keplerian natural theology: the idea that human beings are made in the image of a Mind that is the source of all things.

---

34   Ibid., 72.
35   Keith Ward, *God, Chance & Necessity* (Oxford: Oneworld Publications, 2009), 55.

CHAPTER 13

Image—Human Rationality and the Natural Sciences

What remains of this project is an examination of the third string of Keplerian natural theology: the idea that human beings are made in the image of a rational creator God, an intellectual that kinship makes scientific investigation possible. In the contemporary discipline of philosophy of mind, the Argument from Reason (AR) is used to show that naturalism cannot account for human rationality. According to one version of the AR, there are good reasons to believe that the success of the natural sciences undermines the naturalistic view of the mind; the cognitive capacities crucial to the scientific enterprise, namely, logical and mathematical reasoning, simply would not exist if a reductionist, physics-and-chemistry-only account of the human mind were true. If the AR succeeds, then naturalism cannot provide a reasonable theory of the origin of rational minds, while the Christian doctrine of an immaterial soul made in the image of a rational God seems all the more plausible. It would mean that Kepler's belief in the exclusive resonance between the mind of God and the mind of man has an important ally in contemporary philosophy.

A Brief Intellectual History of the Argument from Reason

The claim that naturalism is incompatible with any account of genuine human rationality has intellectual roots that go back at least as far as the fourth century AD. In *On Free Choice of the Will*, Augustine writes that mankind, unlike inanimate matter and the lower animals, possesses reason and understanding. He describes this capacity as the "eye of the soul" which sets man apart from the rest of creation.[1] He explains that just as we experience bodily sensations from an exclusive, first-person perspective, so our reasoning is only accessible from this same first-person perspective (never by a third party). From the first-person perspective, we are able to actively and freely pursue higher, objective truths (wisdom) with our own minds.[2] Augustine argued that free will plays an essential part in higher reasoning, and he affirmed the existence of objective, immaterial truths such as the laws of logic: "The validity of syllogisms is not something instituted by humans...It is built into the permanent and divinely instituted system of things."[3] As will be seen, these ideas would go on to play key roles in some formulations of the contemporary AR.

Centuries after Augustine, René Descartes (1596–1650) discussed philosophical ideas that also remain relevant to the current discussion. In the *Meditations on First Philosophy*, he laid out his project of epistemology, in which he asks: can human reason give us certainty about anything? In the third meditation, he examines the question of God's existence and whether God desires man to have

---

1 Augustine, *On Free Choice of the Will* (Indianapolis, IN: Hackett Publishing Co., 1993), 41.

2 Ibid., 50.

3 Augustine, *On Christian Teaching*, 59. Boethius, following Augustine, believed that free will is a requirement for a rational nature. See Boethius, *The Consolation of Philosophy*, 99.

reliable rational faculties—both of which he deems necessary: "... if I find that there is a God, I must also inquire whether He may be a deceiver; for without a knowledge of these two truths I do not see that I can ever be certain of anything."[4] Through a chain of reasoning, he concludes that God does exist, has created mankind with an innate idea of Him, and that this knowledge is possible because it is "most probable that in some way [God] has placed His image and similitude" upon human beings.[5] Descartes's implicit assertion seems to be that the rationality of God (Who cannot possibly be a deceiver, because "fraud and deception necessarily proceed from some defect," and God is perfect) and the fact that mankind is made in God's image are the prerequisites for man's veridical reasoning faculties.[6] In the fourth meditation, Descartes relates God-given free judgment with the process of mental deliberation. He says, "I experienced in myself a certain capacity for judging which I have doubtless received from God...as He could not desire to deceive me, it is clear that He has not given me a faculty that will lead me to err if I use it aright."[7] Note that free will is crucial to Descartes's characterization of both divine and human reason— the thing that is free is the thing which thinks. As one scholar puts it, Descartes saw "an indivisible and inalienable freedom—imaging the divine creativeness—at the heart of created rational nature."[8] It is by virtue of this freedom that human beings have the capacity to affirm or deny the truth of a proposition, or to choose to suspend judgment in any instance of rational deliberation.

---

4   René Descartes, *Meditations*, Great Books of the Western World 28 (Chicago: Encyclopaedia Britannica, Inc., 1990), 308.

5   Ibid., 314.

6   Ibid..

7   Ibid., 315.

8   Sophie Berman, "Human Free Will in Anselm and Descartes," *The Saint Anselm Journal* 2.1 (Fall 2004): 1.

Several decades after the publication of Descartes's *Meditations*, John Locke (1632–1704) argued that it is impossible to conceive "that ever bare incogitative matter should produce a thinking intelligent being."[9] In other words, mere material stuff cannot give rise to rationality. In his *Essay Concerning Human Understanding*, he says that one might as well believe that something can come from nothing; just as matter in motion needs a first mover, sentience and rationality must come from a supreme Mind. Even if matter has been in motion from eternity, coalescing into various forms, it could not have produced rational minds:

> But let us suppose motion eternal too; yet matter, incogitative matter and motion, whatever changes it might produce of figure and bulk, *could never produce thought*: knowledge will still be as far beyond the power of motion and matter to produce, as matter is beyond the power of *nothing*, or *non-entity* to produce. And I appeal to everyone's own thoughts, whether he cannot as easily conceive matter produced by *nothing*, as thought to be produced by pure matter, when before there was no such thing as thought, or an intelligent being existing.[10]

For Locke, human rationality must have come from a higher rationality—a mind that is, in some sense, fundamental to the existence of all things. His objective in this section of his essay is to make an argument for theism; he says that because matter does not come from nothing and because rational minds exist and cannot have come from "incogitative matter," the first cause of all things cannot have been purely material and thus non-rational in nature.

A contemporary of Locke, Gottfried Leibniz (1646–1716), developed an ingenious analogy to make a similar point about the blind, mechanistic motions of matter and human rationality. Leib-

---

9 John Locke, *Essay Concerning Human Understanding* (Indianapolis: Hackett Publishing, 1996), 278.

10 Ibid., 279.

niz believed in the existence of a conscious soul, which he referred to as a "monad" or a "simple substance" (meaning something indivisible), that is ontologically distinct from the biological machinery of the human brain. He used what is now known as the famous "mill analogy" to help his readers see that one's mental experiences, which include subjective perceptions of the external world, are not entirely explained by the parts and functions of the brain. When Leibniz refers to a "mill" he means the primitive kind involving a water wheel that turns giant gears which move a grinding stone. He says:

> If we imagine that there is a machine whose structure makes it think, sense, and have perceptions, we could conceive of it enlarged, keeping the same proportions, so that we could enter into it, as one enters into a mill. Assuming that, when inspecting its interior, we will only find parts that push one another, and we will never find anything to explain a perception. And so, we should seek perception in the simple substance and not…in the machine.[11]

Leibniz's point is that we can examine all the "gears" and "wheels" of the brain, but never access the actual *perceptions* occurring in another person's mind. An observer cannot access someone else's mental experiences even though it is possible, in principle, to have complete understanding of the mechanics of the material brain. Leibniz considered this a good reason for believing that a person's mind is not sufficiently explained by matter in motion; there are aspects of the mind that seem to be immune to physical description or third-party empirical investigation.

Thomas Reid (1710–1796), the Scotsman who became known as the Common Sense Philosopher, insisted that it is the mind—not the physical body—that thinks, reasons, and wills. He explained that material entities behave according to the forces acting *upon*

---

11   G. W. Leibniz, *Philosophical Essays* (Indianapolis: Hackett Publishing, 1989), 215.

them, whereas the mind is, by nature, an agent that can act *freely*. For Reid, genuinely rational activity can exist only if the mind has the power to self-direct its train of thought. In other words, if we do not control at least some of our own mental processes, then true rationality, as we commonly define it, does not exist. Moreover, if we are not agents with the power to direct our mental actions, how is it that products of our mind can purposefully conform to external, immaterial standards? Reid says, "No man can believe that Homer's ideas…arranged themselves according to the most perfect rules of epic poetry; and Newton's according to the rules of mathematical composition."[12] Based upon the fact that mental output can follow non-physical standards, such as the stylistic parameters of epic poetry and the laws of mathematics (to use Reid's excellent examples), we should conclude that the outcome of reasoning is not merely the result of a series of blind physical processes that, by definition, can have no conscious objective. Reid, as we shall see, was on to something when he pinpointed the connection between the nature of reasoning and the mind's interaction with abstract objects.

Immanuel Kant (1724–1804), a contemporary of Reid, was another contributor to the discussion. He identified discursive thought—which is marked by analytical reasoning—with the cognitive activity of judgment; he said that "we can reduce all acts of the understanding to judgments, and the *understanding* may therefore be represented as a *faculty of judgment*."[13] Judgment requires what Kant calls *spontaneity*, a conscious mental action that is not instigated by non-rational causes such as deterministic chemical reactions in the brain. Furthermore, as philosopher Henry Allison explains, Kantian judgment can be construed as the act of

---

12 Thomas Reid, *Essays on the Intellectual Powers of Man* (Cambridge: John Bartlett, 1852), 290.

13 Henry E. Allison, "Kant's Refutation of Materialism," *The Monist* 72, 2 (April 1, 1989): 192.

taking something to be such and such.[14] Thus, when one is reasoning through premises to reach a conclusion, the premises are *taken as* (recognized as) justification for the conclusion.[15] Allison explains that "the premises must not only be good or sufficient reasons for asserting the conclusion, they must also be recognized by the reasoner as such."[16] The person carrying out the inferential process must have an awareness of the fact *that* the conclusion follows from the premises as well as *why* the premises are good and sufficient reasons for drawing the conclusion. Note that the recognition that the reasons for accepting a conclusion are good ones involves a mental connection with a standard of reasoning such as a law of logic. Essentially, Kant argues that our reasoning, to be genuine reasoning, must function independently, in accord with "objectively valid normative principles," rather than being the result of a deterministic material causal process.[17] Allison points out that "as long as one links [the premises] with the conclusion in this fashion, one is reasoning."[18] Moreover, all of this process occurs within one single consciousness—an *I* who is self-aware of his or her rational activities. This is referred to in Kantian terms as "unity of consciousness." According to Kant, the material causal model cannot account for these things, and hence, it cannot be the correct explanation for rationality; there is no room in such a model for an *I* and its intentional deliberation.

It may be surprising to some that Charles Darwin (1809–1882), the father of modern evolutionary theory, had doubts about a fully materialist account of the human mind. In his *Origin of Species* (1859), he sought to explain the apparent design in biolog-

---

14 Ibid., 193.
15 Ibid.
16 Ibid.
17 Ibid., 197.
18 Ibid., 193.

ical organisms with a fully naturalistic process—natural selection fueled by random heritable variations. Later, in 1871, he applied his theory to the origin of humans in *Descent of Man*. In *Descent*, he claims that mankind is different only in degree, not in kind, from the rest of the animal kingdom. He dedicated three chapters to his reasons for believing that man's higher cognitive faculties are merely more advanced versions of faculties seen in lower animals, and that human mental powers surely resulted from a steady, upward march of unguided evolutionary change. However, in his autobiographical material, Darwin expressed reservations about the removal of divine agency from man's creation:

> Another source of conviction in the existence of God, connected with the reason and not with the feelings, impresses me as having much more weight [than preceding arguments]. This follows from the extreme difficulty or rather impossibility of conceiving this immense and wonderful universe, including man with his capacity of looking far backwards and far into futurity, as the result of blind chance or necessity. When thus reflecting I feel compelled to look to a First Cause having an intelligent mind in some degree analogous to that of man; and I deserve to be called a Theist.[19]

This is a stunning admission, particularly the remark about man having a mind analogous to a transcendent mind. Darwin goes on to explain how his confidence in this argument for God's existence had waxed and waned over time, ending up weaker than it was when he first penned his *Origin*, yet he expresses concerns about the trustworthiness of the human mind if it is merely an evolved material brain operating according to deterministic processes of cause and effect.[20] In a letter to William Graham dated July 3, 1881,

---

19 Nora Barlow, ed., *The Autobiography of Charles Darwin 1809–1882* (London: W.W. Norton & Company, 1958), 92–93.

20 Ibid., 93.

he writes that "the horrid doubt always arises whether the convictions of man's mind, which has been developed from the mind of the lower animals, are of any value or at all trustworthy. Would any one trust in the convictions of a monkey's mind, if there are any convictions in such a mind?"[21] He concluded, "I cannot pretend to throw the least light on such abstruse problems. The mystery of the beginning of all things is insoluble by us; and I for one must be content to remain an Agnostic."[22] The problem of explaining man's rationality by way of non-rational processes came to be known as "Darwin's Doubt."

In his 1895 book entitled *The Foundations of Belief*, Arthur James Balfour (1848–1930), a revered member of the scientific community who served as British Prime Minister from 1902 until 1905, discusses what he regards as difficulties with the alleged "non-rational origin of reason"—the naturalistic evolutionary emergence of human rationality by way of non-rational causes.[23] If a non-teleological account of the human mind is the case, Balfour argues, then reason came about in the same way and for the same purposes as all our other faculties: survival and reproduction. He recognizes the great explanatory leap between the hypothesis that natural selection could have endowed man with the faculties necessary for biological flourishing and the assumption that it also produced reliable higher rationality:

> For to suppose that a course of development carried out, not with the object of extending knowledge or satisfying curiosity, but solely with that of promoting life, on an area so insignificant as the surface of the earth, between limits of temperature and pressure so narrow, and under general conditions so exceptional,

---

21 Charles Darwin, "To William Graham: 3 July 1881," Darwin Correspondence Project at Cambridge University.

22 Nora Barlow, *The Autobiography of Charles Darwin 1809–1882*, 94.

23 Balfour served as president of the British Association for the Advancement of Science in 1904.

should have ended in supplying us with senses even approximately adequate to the apprehension of Nature in all her complexities, is to believe in a coincidence more astounding than the most audacious novelist has ever employed to cut the knot of some entangled tale. For it must be recollected that the same natural forces which tend to the evolution of organs which are useful tend also to the suppression of organs that are useless.[24]

In short, Balfour saw naturalistic accounts of the origin of rationality as preposterous. During the period of evolution in which our mental capacities were forming, natural selection would have only preserved those that were useful at that time. The aptitude for higher mathematics and advanced logical reasoning would not have conferred an appreciable benefit in the primitive conditions in which our hominid ancestors lived.[25] Balfour goes further, contending that if the naturalistic creed is a true and complete account of human origins, then what we call human reason would be an untrustworthy producer of beliefs: "My beliefs, in so far as they are the result of reasoning at all, are founded on premises produced in the last resort by the 'collision of atoms'…Atoms, having no prejudices in favour of truth, are as likely to turn out wrong premises as right ones; nay, more likely, inasmuch as truth is single and error manifold."[26] In other words, if naturalism is true, then beliefs are determined by matter in motion governed by the laws of physics and chemistry, a process that lacks any tendency towards propositional truth. Why, then, would the resulting beliefs be reliable from the standpoint of truth rather than merely advantageous to survival and reproduction?

---

24 Arthur James Balfour, *The Foundations of Belief* (New York: Longmans, Green, & Co., 1895), 70.

25 Here and later, the broad outlines of alleged evolutionary history are granted for the sake of argument.

26 Ibid., 279.

In the first half of the twentieth century, dramatic progress was made in the natural sciences, and some scientists and mathematicians involved with these advancements participated in the discussion about physical determinism's ramifications for human reason. Consider the words of J. B. S. Haldane (1892–1964), an atheist who was highly educated in both mathematics and biology: "If materialism is true, it seems to me we cannot know that it is true. If my opinions are the result of the chemical processes going on in my brain, they are determined by chemistry, not the laws of logic."[27] In other words, blind chemical processes cannot take principles of good reasoning into account and therefore do not constitute genuine rationality, even though such processes could *accidentally* produce a result that conforms to the laws of logic. (Notice how Reidian he sounds.) In his essay "Possible Worlds" Haldane writes:

> It seems to me immensely unlikely that mind is a mere by-product of matter. For if my mental processes are determined wholly by the motions of atoms in my brain I have no reason to suppose that my beliefs are true. They may be sound chemically, but that does not make them sound logically...In order to escape from this necessity of sawing away the branch on which I am sitting, so to speak, I am compelled to believe that mind is not wholly conditioned by matter.[28]

One key observation about this passage is Haldane's statement about the chemical activity of the brain not making mental content compatible with the external, immaterial laws of logic. He says that when he engages in logical, scientific, and moral thinking, "I am already identifying my mind with an absolute or unconditioned

---

27 J. B. S. Haldane, *The Inequality of Man* (London: Chatto & Windus, 1932), 162.

28 J. B. S. Haldane, *Possible Worlds* (New York: Routledge, 2017), 209. At this point in his career, Haldane considered the possibility of an afterlife. See Endnote 21.

mind."[29] By "unconditioned" he seems to mean a mind that is not subject to the limitations of the physical world. Haldane's tree branch analogy indicates that he regarded any material account of mind as self-defeating, but later in his career, as a result of the development of early computers—machines that could function according to rules of logic and mathematics—Haldane changed his mind about the possibility of such an account. Yet, one could argue that Haldane made a mistake in thinking that brains are like computers, since the latter are able to function successfully only because an intelligent agent has programmed them with specific rules of computational logic. Computers could easily be designed to use faulty logic because there is nothing about their operation that is concerned with *truth*.

Hermann Weyl (1885–1955), a German theoretical physicist and mathematician who contributed to cosmology and quantum mechanics, also saw the problem that the materialist understanding of the mind posed for human rationality. In his book *Mind and Nature*, he says, "When I reason that 2 + 2 = 4, this actual judgment is not forced upon me through blind natural causality (a view which would eliminate thinking as an act for which one can be held answerable) but something purely spiritual enters in: the circumstance that 2 + 2 really equals 4, exercises a determining power over my judgment."[30] Weyl seems to mean that when we reason, we are able to take truth into account and then freely direct our mental processes accordingly. However, if the mind is wholly the result of inevitable physical processes, it simply turns input into output by way of neurochemical reactions that cannot distinguish truth from falsehood.

---

29 Haldane, *Possible Worlds*, 210.

30 Hermann Weyl, *Mind and Nature* (Princeton: Princeton University Press, 2009), 52.

C. S. Lewis's Formulation of the AR

An avid reader of Balfour, C. S. Lewis (1898–1963) made a significant mark in the philosophical debate about the ramifications naturalism has for human reason.[31] In fact, his name is often directly linked with the AR. In the final (post-Anscombe) version of his argument, which was published in the second edition of his book *Miracles*, Lewis said,

> Obviously many things will only be explained when the sciences have made further progress. But if Naturalism is to be accepted we have a right to demand that every single thing should be such that we see, in general, how it could be explained in terms of the Total System. If any one thing exists which is of such a kind that we see in advance the impossibility of ever giving it *that kind* of explanation, then Naturalism would be in ruins…For by Naturalism we mean the doctrine that only Nature—the whole interlocked system—exists.[32]

By Lewis's lights, the phenomenon of human rationality is precisely the kind of thing that cannot have a naturalistic explanation, no matter what degree of progress is eventually made in the sciences. In other words, rationality defies naturalism *in principle*, because of the kind of thing it is; attempting a matter-and-energy, laws of physics and chemistry explanation is an exercise in futility, for such an account would eliminate the very essence of human reason. This would hold, he says, no matter what actually goes on at the particle level of material reality (determinism or indeterminism, to use contemporary terms), since prediction of future physical states

---

31 C. S. Lewis listed Balfour's *Theism and Humanism* as one of the top ten books that had influenced him intellectually. He gave his top ten most influential books list in a 1962 interview by *The Christian Century* magazine.

32 C. S. Lewis, *Miracles* (New York: HarperCollins, 2015), 17. For details on Anscombe, see Endnote 22.

is still reliable at the macro level due to the law of averages. Any naturalistic account of human reason entirely lacks teleology of any sort; it boils down to blind matter in motion. However, rationality (as we ordinarily understand it) gives us true insight into realities that exist beyond our own minds. In other words, it enables us to arrive at true propositions about the world around us as well as those that exist in the abstract, such as mathematical and logical truths—which have no power of causation in the material realm.

In his articulation of the AR, Lewis draws a key distinction between the types of reasons (the *because*) that can be given to account for beliefs. What he calls a *Cause and Effect* reason for a belief involves "a dynamic connection between events or 'states of affairs,'" while the *Ground and Consequent* reason for a belief indicates "a logical relation between beliefs or assertions."[33] Lewis insists that genuinely rational thought necessarily includes a Ground and Consequent relationship between each successive step in a chain of reasoning. Also, thought A does not cause thought B merely by *being* a ground for it; A causes B because the thinker *sees that* B logically follows from A (recall that Kant made a similar point). Lewis calls this a first-person "act of knowledge." If naturalism is true, however, all thoughts are nothing more than the result of Cause and Effect alone; the set of neurons firing during thought A are the physical cause of the subsequent neuron firing pattern that produces thought B. Any Ground and Consequent relationship is decidedly irrelevant, playing no part whatsoever in the production of thought B. Lewis asks, "How could such a trifle as lack of logical grounds prevent the belief's occurrence or how could the existence of grounds promote it?"[34] Reasons (grounds) are immaterial entities that cannot impact the physical stuff of the brain.

---

33   C. S. Lewis, *Miracles*, 22.

34   Ibid., 24.

A serious problem with naturalism, Lewis says, is that it "offers what professes to be a full account of our mental behavior; but this account, on inspection, leaves no room for the acts of knowing or insight on which the whole value of our thinking, as a means to truth, depends."[35] What is needed in reasoning is genuine Ground and Consequent causation involving direct conscious awareness of the logical relationships between one mental event and the mental event that follows as a result. Lewis goes on to explain that acts of thinking (mental events) are a very particular sort of event in that they have *aboutness*, they have informational content and can be true or false. Physical events such as physiological brain states lack this character; they are not *about* anything, nor can they be classified as true or false.

Like Balfour, Lewis was convinced that a blind evolutionary process driven by natural selection cannot explain true rationality. Such a process merely promotes survival and reproduction; therefore, the biological hardware of thinking would have evolved to enhance an organism's physical fitness, not the factual reliability or logical validity of its mental activities. He says that "it is not conceivable that any improvement of responses [of an organism to its environment] could turn them into acts of insight, or even remotely tend to do so. The relation between response and stimulus is utterly different from that between knowledge and the truth known."[36] In the scientific enterprise, he says, "knowledge is achieved by experiments and inferences from them" not a biochemical refinement of the scientist's response to sensory observations.[37] It is relevant and quite significant that elsewhere Lewis writes that the "scientific point of view" (meaning naturalism) cannot even account for science itself.[38] What he means is that natu-

---

35   Ibid, 27.

36   Ibid., 28.

37   Ibid.

38   C. S. Lewis, *The Weight of Glory* (New York: HarperOne, 2000), 140.

ralism explains away the genuine rationality needed for scientific inference; thus to legitimately value scientific practice should be to discard any view that damages its own epistemic foundation.

In addition to outlining these formidable problems for a naturalistic account of rationality, Lewis went as far as to say that any such explanation is doomed to be self-defeating. As Stewart Goetz explains:

> Lewis concluded that naturalism…must be false because reasoning (making inferences) violates the causal closure of the material world according to which the explanation of a mental event can include nothing other than what is material in nature. And if naturalism is itself a philosophical position that is arrived at on the basis of reasoning, then the game is up. Naturalism ends up being self-defeating because those who espouse it typically arrive at it on the basis of reasoning.[39]

The conclusion that naturalism is a correct account of reality is one that requires reasoning, but if that process is nothing more than the inevitable outcome of blind biological processes taking place within a causally closed system (i.e., a system that is closed to immaterial agency), *there is no reason to believe that conclusion.* One may be reminded of Haldane's remark about sawing off the tree branch one is sitting upon. Lewis quotes Haldane's words about motions and atoms in the brain, and argues that if naturalism is correct, "It discredits our processes of reasoning or at least reduces their credit to such a humble level that it can no longer support Naturalism itself."[40] The theist, however, bears none of the aforementioned difficulties when it comes to explaining the origin and reliability of human reason. Lewis says, "For him, reason—the reason of God—is older than Nature, and from it the orderliness of Nature, which alone enables us to know her, is derived. For

---

39 Stewart Goetz, "The Argument from Reason," *Philosophia Christi* 15, no. 1 (2013): 50.
40 Lewis, *Miracles*, 22.

him, the human mind in the act of knowing is illuminated by the Divine reason. It is set free, in the measure required, from the huge nexus of non-rational causation; free from this to be determined by the truth known."[41] If human beings are creatures with immaterial souls made in the image of a rational God, a way is opened for genuine rationality to enter the world picture.

The Lewisian AR in Contemporary Philosophy

The AR continues to be actively discussed in contemporary philosophical literature, and although it has been developed into several different versions, the Lewisian AR is still a significant part of the ongoing conversation. Victor Reppert, who has done extensive work in this area, explains that Lewis "does not answer every question that a critic might ask" nor does he offer "a set of arguments sufficiently polished to persuade persons with technical training in philosophy or other disciplines."[42] Thus, one must explore the different ways in which Lewis's thought might be developed and defended—which Reppert and other contemporary philosophers have indeed done.

Goetz interprets Lewis's AR as a deductive argument by which we can conclude the falsehood of naturalism:

1) If naturalism is true then we do not reason.

2) We reason.

3) Naturalism is false.[43]

---

41 Ibid., 34.

42 Victor Reppert, *C.S. Lewis's Dangerous Idea: In Defense of the Argument from Reason* (Downers Grove: InterVarsity Press, 2003), Kindle location 59.

43 Goetz, "The Argument from Reason," 51.

In this version of the AR, the fact that human reasoning is real (premise 2) rather than illusory is taken as a given that (according to premise 1) cannot be accounted for under the terms of naturalism. To argue that we do not reason is ultimately a self-defeating exercise because argumentation utilizes reasoning. Rationality includes *aboutness* (what philosophers call intentionality) of mental events and the reality of Ground and Consequent reasons for beliefs which, as Lewis argued, are incompatible with causal closure of the material world (a central dogma of naturalism). According to Lewis, there is no room in naturalism for the mental causation of beliefs; all mental phenomena are, or are the direct result of, physical brain states. Without mental causation, there would be no authentic rationality; beliefs would be fully determined by physical processes (whether determinate or indeterminate at the quantum level). The laws of logic would have no bearing on the beliefs we come to hold; we would not actually take any facts into account when we deliberate towards a conclusion; we would not weigh evidence and draw inferences; and we would not make judgments. Such things would be illusory.

Peter van Inwagen, though not a naturalist himself, does not believe that the Lewisian AR succeeds in its goal.[44] Specifically, van Inwagen objects to the claim that the physical determination of beliefs necessarily excludes true rational grounding of those same beliefs. If a belief can have both physically determined causes and rational grounding at the same time, he argues, then Lewis's AR fails to *show* that according to naturalism, we do not reason. Lewis, in other words, has not demonstrated such *exclusion* to be the case and thus has not provided a so-called "cardinal difficulty" for naturalism. In response to van Inwagen, Todd Buras and Brandon Rickabaugh say that something called the problem of *causal drainage* fortifies the Lewisian AR against van Inwagen's objection. They

---

44 See Peter van Inwagen, "C.S. Lewis' Argument Against Naturalism," *The Journal of Inkling Studies* 1, no. 2 (October 2011): 25–40.

contend that material-mechanistic explanations of any phenomenon—in this case, a belief—are considered, according to naturalism, to be "in principle complete, that is, sufficient to explain everything that needs explaining."[45] There is no work left to be done by the Ground and Consequent relation in forming the belief in question because its role is "drained away" by the sufficiency of the material-mechanistic explanation for the belief. "With respect to belief," Buras and Rickabaugh argue, "naturalism entails that any belief b, if it is to be explained, will in principle be given a sufficient mechanistic explanation, such that non-mechanistic explanations are eliminable. That is, if [the] mechanistic explanation is complete, personal [Ground and Consequent] explanations are eliminable."[46] Goetz agrees; he asks: "…how does the mental cause do any genuine explanatory work? It seems that the material cause preempts the mental cause and makes the latter explanatorily impotent."[47] As Roger Trigg puts it, "Questions of justification of belief, and of its rational grounding, are squeezed out."[48]

Another problem for completely mechanistic explanations for beliefs is that they make the aboutness of antecedent beliefs superfluous in explaining why someone comes to hold a subsequent belief. Fred Dretske, a materialist, offers the following analogy regarding the superfluity of the alleged aboutness (what he calls the "semantic properties") of a material cause (a brain event) and its effect (a resulting belief or behavior):

> Meaningful sounds, if they occur at the right pitch and amplitude, can shatter glass, but the fact that these sounds have a meaning is surely irrelevant to their having this effect. The glass would

---

45 Brandon Rickabaugh and Todd Buras, "The Argument from Reason, and Mental Causal Drainage: A Reply to Peter van Inwagen," *Philosophia Christi* 19, no. 2 (2017): 385.

46 Ibid., 389.

47 Goetz, "The Argument from Reason," 62.

48 Roger Trigg, *Beyond Matter*, 64.

shatter if the sounds meant something completely different or if they meant nothing at all. This fact doesn't imply that the sounds don't have a meaning, but it does imply that their having meaning doesn't help explain their effects on the glass...If the semantic properties of reasons, the what-it-is we believe and desire, is irrelevant to explaining their causal properties...then the fact that they are causes, taken by itself, is or should be very little solace indeed. For it leaves us without the resources for understanding the explanatory role of reasons...[49]

In the case of reasoning, it doesn't matter what the belief preceding a conclusion is *about*, because its content has no power to cause that subsequent belief. The what-it-is we believe, as Dretske puts it, does not matter in the least; the same subsequent belief would still arise as a result of the material processes at work. In other words, initial *mental* states (beliefs) do not have a causal role to play because the physical brain states on which they allegedly supervene fully account for the resulting brain states and the corresponding mental states (the subsequent beliefs). For example, my belief that the weatherman is correct about an impending downpour (belief A) has nothing to do with my subsequent belief that I should take my umbrella with me when I leave my house (belief B). Instead, one blind brain state led to another, and the fact that I have a reason (A) for the belief that I should take my umbrella with me (B) is essentially irrelevant in any causal sense whatsoever. Buras and Rickabaugh summarize the issue this way:

> The problem is not that one lacks reasons to believe. The problem is that naturalism leaves no room for one's reasons to play a causal role in the production of belief. If mental states are not causally involved in the production of other mental states, it fol-

---

49 Fred Dretske, "Reasons and Causes," *Philosophical Perspectives* 3 (1989), 3.

lows that no one believes anything based on reasons, including the belief that naturalism is true.[50]

Recall that Lewis himself was aware of this problem. He said that if a belief "is simply one link in a causal chain which stretches back to the beginning and forward to the end of time," then "How could a trifle as lack of logical grounds prevent the belief's occurrence or how could the existence of grounds promote it?"[51] Lewis did not take causation out of the equation, he just denied that causation is physical where rationality is concerned; the cause is not part of nature, what he calls "the great mindless interlocking event."[52] J. P. Moreland, a preeminent philosopher of mind, highlights a related problem: "If we reject the notion that there cannot be two sufficient efficient causes for some physical event that is, in fact, caused, and grant for the sake of argument that there is a mental cause for some physical event, then we have two competing causal stories."[53] The point is, both causal stories cannot be true.

Buras and Rickabaugh continue: "Of course, naturalism might explain how mechanical devices like calculators and computers act in accordance with reason. However, naturalism cannot explain how persons act from reason, where reason itself is the explanation."[54] Machines can indeed be programmed to calculate data according to the rational laws of mathematics or other rules, but this is not the same thing as having inherent rationality, which requires consciously using logical parameters to reason to a conclusion—Lewis's Ground and Consequent causation. The computer

---

50 Rickabaugh and Buras, "The Argument from Reason, and Mental Causal Drainage: A Reply to Peter van Inwagen," 397.

51 Lewis, *Miracles*, 24.

52 Ibid., 10.

53 J. P. Moreland, *Consciousness and the Existence of God* (New York: Routledge, 2009), 17–18.

54 Rickabaugh and Buras, "The Argument from Reason, and Mental Causal Drainage: A Reply to Peter van Inwagen," 397.

gives mechanically determined output; it cannot *see that* there are grounds for the consequent, thus it does not truly reason. Similarly, under the naturalistic paradigm, the material brain is merely a biochemical entity, and thus its output cannot be deemed the product of genuine rationality even if said output is *consistent* with reason.

CHAPTER 14

The Success of the Natural Sciences as Evidence for Authentic Rationality

As previously discussed, the question of whether humans genuinely reason has a close and strong connection with the existence and success of the natural sciences. Roger Trigg explains, "Rationality and the human freedom to exercise it make scientific investigation and argument possible."[1] But, he argues, if everything, including the activity of the human brain, is entirely the result of law-governed material processes, "No room is left for the exercise of reason. The kind of reality envisaged by physicalism cannot allow for the independent reality of mind, belief, or reason. All is in physical form and physically determined."[2] The problem this poses for the scientist who holds to physicalism (and is thus a naturalist) is inescapable; he must come to grips with the fact that science entirely depends upon the core assumption that true rationality exists, yet that assumption cannot be supported by the naturalist's own worldview. Thus, there is a direct connection between the Lewisian AR and scientific activity. Taking the foregoing discussion into account, an expansion on the Lewisian AR might be framed as follows:

---

1 Roger Trigg, *Beyond Matter*, 71.

2 Ibid., 119–120. Physicalism, as Trigg defines it, is the position that nothing exists besides the material stuff of the universe.

1. The success of the natural sciences heavily depends upon the ability to freely draw logical inferences and carry out mathematical deliberation—activities that require genuine human rationality.
2. Genuine human rationality would not exist if a naturalistic account of the human mind were correct.
3. Therefore, the observable enormous success of the natural sciences strongly undermines a naturalistic account of the human mind.

Premise (1) will here be treated as a matter of fact. Premise (2) will be expanded and defended to complete the case for the conclusion (3).

Genuine human rationality includes at least four different issues that pose seemingly intractable problems for naturalism: 1) intentionality of mental content, 2) mental causation, 3) free agency, and 4) cognitive interaction with abstract objects (numbers, mathematical truths, laws of logic, etc.). Each of these will be examined in turn.

A naturalistic account of human reason runs head-on into the problem of intentionality—the aboutness of mental events. When we reason, we are interacting with informational content such as meanings, concepts, logical relations, and epistemic relations, none of which are material causal entities.[3] The problem is, physical brain events that coincide with our first-person conscious awareness of mental content cannot be said to be *about* anything at all; brain anatomy and physiology involve arrangements and dynamics of matter that simply *are*. A physical entity such as a network of firing neurons can no more be *about* anything than a tree, a galaxy, or a grain of sand can be *about* anything. Physical objects do not have the sort of informational content that thoughts do, even though

---

3 J. P. Moreland, *The Recalcitrant Imago Dei* (London: SCM Press, 2009), 87.

physical media can be used to *carry* information that is encoded by an independent mind.

Consider the fact that a mathematical proposition (which involves concepts and their relations) can be true or false while a brain state (matter in motion) cannot be said to be true or false. Suppose a person is having the train of thought, "Three times three is nine; therefore, the square root of nine is three," in response to a question on a math exam. This mental content is true, but the brain state that is occurring simultaneously is merely physics and chemistry, which cannot be described as true or false. Also, one's private thoughts about nine, three, and their mathematical relationships are available only to the thinker, introspectively, while the physical brain activity associated with these thoughts is, in principle, available to a third party, such as a neuroscientist utilizing monitoring equipment. The point is that the mental events and the correlating brain events do not have the same properties, and thus it would be incorrect to say that mental events simply *are* brain events.[4]

The challenge for the proponent of naturalism is to come up with a feasible account of intentionality that stays within the confines of their philosophy of nature. Philosopher and cognitive scientist Daniel Dennett has proposed that our first-person conscious experience of mental content is merely a "user-illusion" that has resulted from naturalistic evolutionary processes. For Dennett, our brains are analogous to computers, and somewhere along the meandering pathway of evolution, random mutations flipped the switch that brought about consciousness. This was the watershed event that made sensory perceptions of the external world and first-person introspection (which is required for rationality) possible. In essence, Dennett's tactic is to simply get rid of the problem by claiming that phenomena such as the intentionality of mental content and mental causation are illusory byproducts of neuro-

---

4   This is an application of Leibniz's Law.

chemical activity. He argues that what we think of as comprehension of the world is nothing more than the evolutionary (both biological and cultural) accumulation of various competences. He says, "Our thinking is enabled by the installation of a virtual machine made of virtual machines made of virtual machines...When we evolved into an us, a communicating community of organisms that can compare notes, we became the beneficiaries of a system of user-illusions that rendered versions of our cognitive processes—otherwise as imperceptible as our metabolic processes—accessible to us for purposes of communication."[5] Ultimately, Dennett's assertions seem to amount to a self-defeating argument; if humans cannot get "outside" of the user-illusions that characterize mental life, how could one possibly know that Dennett's explanation is an objectively true account of what we call rationality?

As C. S. Lewis noted—and Victor Reppert emphasizes—something else that is needed to preserve authentic rationality according to naturalism is an account of causation in which one mental event causes another mental event by virtue of content and logical relations: a process known in contemporary philosophy as mental-to-mental causation.[6] In an attempt to provide such an account, some naturalists espouse a non-reductive form of physicalism in which mental events "emerge" from underlying physical brain events and these emergent mental events can actually *cause* subsequent mental events. In this theoretical scenario, a mental event M emerges from a physical brain event P and then causes the next physical brain event P^* from which the next mental event M^* emerges. M^* is said to be caused by M, even though both have a physical substrate upon which they entirely depend for their existence (P^* and P, respectively). According to this view, it remains the

---

5 Daniel Dennett, *From Bacteria to Bach and Back: The Evolution of Minds* (New York: W.W. Norton & Company, 2017), 341, 344.

6 Victor Reppert offers this argument in syllogistic form in the fourth chapter of his book, *C.S. Lewis's Dangerous Idea: In Defense of the Argument from Reason*.

case that physical substrate P has a causal relationship with physical substrate P^* from which M^* emerges. Thus, for M to be the true cause of M^*, it must act *top-down*, causing P^* from which M^* emerges. As Jaegwon Kim explains, "the only way to save the claim that M caused M^* appears to be to say that M caused M^* by *causing* P^*. It makes sense to think that in order to bring about an emergent phenomenon (or a supervenient property), you must bring about an appropriate basal condition from which it will emerge."[7] This raises the question of why M would not be superfluous in a situation where P^* is already caused by P, and M^* emerges from P^* no matter what is going on with M. Kim asks:

> Why cannot P do all the work in explaining why any alleged effect of M occurred? If causation is understood as nomological (law-based) sufficiency, P, as M's emergence base, is nomologically sufficient for it, and M, as P^*'s cause, is nomologically sufficient for P^*. It follows that P is nomologically sufficient for P^* and hence qualifies as its cause...Moreover, it is not possible to view the situation as involving a causal chain from P to P^* with M as an intermediate causal link. The reason is that the emergence relation from P to M cannot properly be viewed as causal. This appears to make the emergent property M otiose and dispensable as a cause for P^*; it seems that we can explain the occurrence of P^* simply in terms of P, without invoking M at all. If M is to be retained as a cause of P^*, a positive argument has to be provided.[8]

It is precisely a good argument for M as a cause of P^* that is missing. Also, as Kim points out, any such argument would not avoid the problem of overdetermination, since P is a sufficient cause of P^*, and M^* would emerge from P^* even if M did not exist. This presents a serious problem for emergentism which, to be a successful account of human rationality, must include a non-super-

---

7 Jaegwon Kim, "Emergence: Core ideas and issues," *Synthese* 151 (2006): 557.

8 Ibid., 558.

fluous role for downward causation from M to P^*, not to mention a plausible explanation for how an emergent, immaterial mental event can have causative power in the physical realm. Kim does not believe that a successful account of either has been provided.

It seems correct to say that emergentism fails in providing an explanation for how M could make any difference whatsoever in our thought processes. The first premise of Stewart Goetz's formulation of the Lewisian AR seems incredibly resilient: If naturalism is true, then we do not reason. Mental events are physically determined, and physical determination is unaffected by the informational content and logical relations of the emergent mental events associated with them; in short, *mental events are impotent epiphenomena*. However, based on the success of the natural sciences, we know that humans *do* indeed reason, and often reason extremely well, which seems to undercut the naturalistic position.

Some naturalists bite the bullet and admit that their conception of the mind rules out free will and thus they conclude that our intuitive belief that we have free will is illusory. For instance, in his book *The Illusion of Conscious Will*, Daniel Wegner writes:

> Yes, we feel that we consciously cause what we do, and yes, our actions happen to us. Rather than opposites, conscious will and psychological determinism can be friends. Such friendship comes from realizing that the feeling of conscious will, then, may involve exploring how the mechanisms of the human mind create the experience of will. And the experience of conscious will that is created in this way need not be a mere epiphenomenon. Rather than a ghost in the machine, the experience of conscious will is a feeling that helps us to appreciate and remember our authorship of the things our minds and bodies do.[9]

For Wegner, we may be the *authors* of what our minds do by virtue of being the physical possessor of our brains, but our minds do

---

9 Daniel Wegner, *The Illusion of Conscious Will* (Cambridge, MA: MIT, 2002), ix.

what they do deterministically. Sounding much like Dennett, he says, "The experience of will, then, is the way our minds portray their operations to us, not their actual operation."[10] But, to "appreciate and remember our authorship" in this way is not to salvage free will at all. If the Lewisian AR is correct, then a naturalistic account of the human mind excludes free will and thus rules out rationality altogether.

When we mentally deliberate, we must be free to carry out the process according to external parameters and then choose a reasons-based conclusion. When presented with a set of facts about which we wish to draw a logical inference, we must be able to make free choices during our mental deliberation, taking into account all sorts of information along the way and reconciling that information with what we already know to be true. This step-by-step practice includes making unfettered judgments about the observations in question and drawing appropriate connections to known facts using the relevant rules of good reasoning. However, if the naturalistic account is true, all our brain activity is beyond our conscious control and the necessary free agency for such deliberation does not exist; all mental activities are determined at the physical level. J. P. Moreland explains:

> Acts of deliberation presuppose that the rational process is 'up to me' and is not determined prior to or during the process. The conclusion is drawn freely. My act of deliberation itself contributes by way of exercises of active power to what outcome is reached. Acts of deliberation presuppose that there is more than one possible conclusion one could reach, but if determinism is true, there is only one outcome possible, and it was fixed prior to the act of deliberation by forces outside the agent's control.[11]

---

10 Ibid., 96.

11 Moreland, *The Recalcitrant Imago Dei*, 74.

Moreland goes on to say that when we deliberate, we analyze the available evidence, considering the weight of each piece as well as how well each implies a certain conclusion, which requires the ability to freely form judgments. If we were physically compelled in this activity by deterministic chemical processes in the brain, we could not be said to be reasoning at all. Moreover, if there is a certain conclusion that we *ought* to draw given the information we have and the laws of logical inference, this assumes that we *can* freely draw that conclusion.[12]

What, then, of science? Trigg says,

> Science assumes as a basis for its own existence the presence of a human rationality that rises above the linkages between cause and effect. The production of science cannot itself be part of a normal causal process...The practice of science depends on weighing arguments to see how strong the evidence is and even to decide what is to count as evidence or what is irrelevant to the matter in hand.[13]

Thus, it is quite difficult to see how a naturalistic conception of human reason could ever explain the sciences, which depend upon free mental deliberation in acts of mathematical and logical reasoning. In scientific investigation, the scientist must be both free to make decisions as he or she proceeds down the pathway of discovery and able to freely accept or resist a conclusion to which initial evidence seems to lead. An interesting case in point is what became known as Albert Einstein's "little lamb."

In 1915, when Einstein presented his paper on general relativity, his extraordinarily complex mathematical theory contained a fudge factor he called the cosmological constant, represented by the Greek letter *lambda* (Λ). The purpose of inserting this constant into his calculations was to avoid the philosophically distasteful

---

12 Moreland, *The Recalcitrant Imago Dei*, 74.

13 Trigg, *Beyond Matter*, 64–65.

(for Einstein) implication of the equations: a non-static universe (a universe that has changed in size over time). In developing his theory, Einstein demonstrated the rational capacity to manipulate abstract concepts in such a way that they mapped onto known or assumed facts about the material world, and to recognize and correct any internal mathematical incoherence in the theory. When he realized that the equations were implying a cosmic history he rejected on *philosophical* grounds, Einstein devised an *ad hoc* mathematical patch that made general relativity compatible with the idea of an eternally static universe. He clung tenaciously to his cosmological constant for quite some time, despite objections from other theoretical physicists whose solutions to the equations indicated an expanding universe. According to some accounts, journalists joked that everywhere that Einstein went, his "lamb" was sure to go. Eventually, empirical data that strongly supported the expansion of space-time confirmed Einstein's blunder, and he modified his theory. What this account suggests is that advanced scientific rationality requires free agency: the ability to purposefully assemble and manipulate mathematical abstractions into a coherent theory, appropriately apply these mathematical constructs to the physical world, and make decisions along the way based upon any theoretical insights or philosophical ramifications that arise. Clearly, these would be impossible if brain activity were fully determined by the blind laws of physics and chemistry.

Einstein developed mathematical equations to describe the universe, *saw that* his equations led to a physical state of affairs he was not satisfied with, and then modified them to achieve the desired implications. This seems to defy, in principle, any naturalistic explanation of advanced scientific reasoning (reasoning that goes far beyond drawing basic generalizations from observed cause and effect). Besides free mental deliberation, something else that needs to be explained is the human mind's ability to apprehend immaterial truths (including those of mathematics and logic) and then deliberate towards conclusions using those truths along with the universal rules of correct reasoning. How is it, according to naturalism, that we could ever *see that* a proposition of logic—such as the law of non-contradiction—is inevitably, always and everywhere true? Atheist philosopher Thomas Nagel asks, "What is the faculty that enables us to escape from the world of appearance presented by our prereflective innate dispositions, into the world of objective reality?"[14] We seem to grasp abstractions such as mathematical and logical truths in a different way than truths gleaned through sensory experience:

> We reject a contradiction just because we see that it is impossible, and we accept a logical entailment just because we see that it is necessarily true. In ordinary perception, we are like mechanisms governed by a (roughly) truth-preserving algorithm. But when we reason, we are like a mechanism that can see that the algorithm it follows is truth-preserving.[15] Something has happened that has gotten our minds into immediate contact with the rational order of the world, or at least with the basic elements of that order, which can in turn be used to reach a great deal more. That enables us to possess concepts that display the compatibility or incompatibility of particular beliefs with general hypotheses.[16]

Indeed, it seems that this is precisely the kind of situation Einstein faced when he realized that the theory of general relativity could not, without factoring in the cosmological constant he concocted, be logically consistent with his static conception of the cosmos. Nagel recognizes that there must be some sort of deep connection between the human mind and the inherent rationality of nature, yet his own worldview is too narrow to account for such a thing.

---

14 Thomas Nagel, *Mind and Cosmos: Why the Materialist Neo-Darwinian Conception of Nature is Almost Certainly False* (New York: Oxford University Press, 2012), 78–79.

15 Here, Nagel follows Kant and Lewis.

16 Nagel, *Mind and Cosmos*, 80.

Another challenge for naturalism is explaining how a physical brain event could come to have informational content that aligns with abstract mathematical laws. Such laws cannot have a sensory impact that triggers a brain event. Moreland has argued that if our minds are merely physical, there can be no explanation for how they have come to know such things, because "to know mathematical truths, [one] must have a soul or mind that has the innate power to grasp non-physical, abstract objects."[17] Otherwise, how would a physical brain come to have knowledge of something (the abstract object) that cannot impact it on a physical level? Moreland classifies abstracta such as mathematical truths as *a priori* knowledge, which is causally inert. "Even if one tries to specify a causal or quasi-causal relation in which [abstract objects] stand to epistemic subjects," he says, "that specification will go far beyond the resources of naturalism."[18] Similarly, Angus Menuge has argued that the physical causal closure of the world excludes the possibility of abstracta having any effect on the human brain, but theists (who have the option of rejecting causal closure) do not face the same problem, because he or she "can maintain...that the world includes immaterial substance, such as God and souls, that are capable of transcending the limitations of the material world, and whose thoughts can access abstract truths."[19] Alvin Plantinga remarks that "it seems sensible to think that a necessary condition of our knowing about an object or kind of object is our standing in some kind of causal relation to that object or kind of object. If this is so, however, and if, furthermore, numbers and their kin are abstract objects, then it looks as though we couldn't know any-

---

[17] Quoted from a private email correspondence with J. P. Moreland. This harmonizes exceedingly well with Kepler's idea about God structuring the human mind to grasp mathematics.

[18] Moreland, *The Recalcitrant Imago Dei*, 78.

[19] Angus Menuge, "Knowledge of Abstracta: A challenge to Materialism," *Philosophia Christi* 18, no. 1 (2016): 9–10.

thing about them."[20] If naturalism is true, and all brain events are determined by prior brain events and/or physical stimuli from the environment, there seems to be no "in" for abstract truths to play a causal role in one concluding (to use the earlier example) that the square root of nine is three. Brian Leftow explains how the situation with knowledge of a necessary truth (P) is different according to theism: "Knowing that P and wanting us to know it, God brings it about that we believe that P—say, by hardwiring us innately to do so given suitable thought-experiments…For God has cognitive contact with the truth's ontology, this conditions His action and His action leads to our beliefs."[21] The phrase "hardwiring us innately" can be taken to be part of the *imago Dei* such that there is a rational kinship between Creator and creature. This is precisely what Keplerian natural theology claims.

Can Naturalistic Evolution Solve the Problem?

It is important to note the subtle distinction between the questions of whether naturalism can account for the nature of rationality and whether naturalistic evolutionary processes are sufficient to explain how advanced cognitive faculties came about. To be sure, each question must be answered in the same way (both affirmative or both negative), but if the very nature of rationality defies naturalistic explanation, then it seems that the AR succeeds and evolutionary neurobiology is rendered irrelevant. However, it is worth briefly examining an additional difficulty for naturalism that arises from the theory of evolutionary emergence.

Some naturalists dismiss the AR on the grounds that, by their lights, a plausible neo-Darwinian account can be given for the

---

20   Alvin Plantinga, *Where the Conflict Really Lies*, 291.

21   Brian Leftow, *God and Necessity*, 74.

emergence of higher reasoning, even if some of the details—such as mental causation, free agency, and the intentionality of mental content—have yet to be fully explained. They gloss over a serious weakness in the naturalistic story: how random mutations feeding the engine of natural selection could produce such a superfluously advanced form of rationality (recall that Balfour made this very point over a century ago). According to the neo-Darwinian paradigm, natural selection preserves traits that contribute to survival and reproduction and weeds out those that are harmful or that provide no significant advantage. Consider the cognitive abilities that are involved in scientific work, particularly the ability to grasp the complex systems of mathematics used in fields such as theoretical physics. This aptitude for higher abstract mathematics has no discernible survival value in the environment that our ancestors inhabited. As Plantinga says:

> Current physics with its ubiquitous partial differential equations (not to mention relativity theory with its tensors, quantum mechanics with its non-Abelian group theory and current set theory with its daunting complexities) involves mathematics of great depth, requiring cognitive powers going enormously beyond what is required for survival and reproduction. Indeed, it is only the occasional assistant professor of mathematics or logic who needs to be able to prove Godel's first incompleteness theorem in order to survive and reproduce.[22]

The problem Plantinga underscores here is how blind processes of evolution by natural selection would ever produce or preserve the advanced cognitive capacities he describes. Moreover, such an impressive mathematical aptitude has no apparent sexual selective advantage either: "What prehistoric female would be interested in a male who wanted to think about whether a set could be equal in

---

22 Plantinga, *Where the Conflict Really Lies*, 286.

cardinality to its power set, instead of where to look for game?"[23] The currently popular response—that this higher rationality is a "spandrel," a spin-off of the evolutionary process—"sounds pretty flimsy, and the easy and universal availability of such explanations makes them wholly implausible" says Plantinga.[24] Physicist and agnostic Paul Davies concurs with Plantinga's assessment; he says:

> One of the oddities of human intelligence is that its level of advancement seems like a case of overkill. While a modicum of intelligence does have a good survival value, it is far from clear how such qualities as the ability to do advanced mathematics... ever evolved by natural selection. These higher intellectual functions are a world away from survival "in the jungle." Many of them were manifested explicitly only recently, long after man had become the dominant mammal and had secured a stable ecological niche.[25]

In a later book, Davies riffs on the same theme while still avoiding a theistic explanation:

> Mindless, blundering atoms have conspired to make not just life, not just mind, but understanding. The evolving cosmos has spawned beings who are able not merely to watch the show, but to unravel the plot...Through science and mathematics, we not only observe the drama of nature, but we have been able, albeit only partially so far...to glimpse the deep, hidden subtext of nature in the form of its subtle mathematical laws and principles, and to gain some understanding of how the universe is put together and works as a coherent system.[26]

---

23   Plantinga, *Where the Conflict Really Lies*, 287.

24   Ibid.

25   Paul Davies, *Are We Alone?* (New York: Orion Productions, 1995), 85.

26   Paul Davies, *The Goldilocks Enigma: Why is the Universe Just Right for Life?* (Boston: Houghton Mifflin, 2008), 5, 255.

Again, it seems like quite a stretch to credit natural selection acting upon random genetic mutations for the production of minds capable of the complex intellectual exercises required for the existence and advancement of the natural sciences.

Science, Therefore Incorporeal Minds

Recall that Goetz's formulation of the Lewisian AR goes as follows: If naturalism is true we do not reason; we reason; thus, naturalism is false. Considering the nature of human reason and naturalism's inability to account for it, the Lewisian AR seems quite compelling. By applying the AR to the obvious successes of the natural sciences (which require genuine human rationality), we may quite reasonably conclude that these successes constitute a defeater of naturalism. We might call this "the argument from scientific success." To recap:

1. The success of the natural sciences heavily depends upon the ability to freely draw logical inferences and carry out mathematical deliberation, activities that require genuine human rationality.

2. Genuine human rationality would not exist if a naturalistic account of the human mind were correct.

3. Therefore, the observable grand success of the natural sciences strongly undermines a naturalistic account of the human mind.

What, then, is the best alternative? If naturalism's dogma of causal closure is false and mental activity is not fully determined by physical brain states, then it stands to reason that the human mind is, or at least involves, a conscious, insightful, immaterial agent with the free will necessary for rational deliberation. It seems, then, that the traditional doctrine of the immaterial human soul made in the image of God (the third string of Keplerian natural theology) is on good philosophical footing.

Several prominent Christian thinkers agree that the existence of science carries philosophical and theological implications. John Lennox, a contemporary Oxford mathematician and theologian, has made remarks in lectures, interviews, and his written work about how the comprehensibility of nature points to both theism and the doctrine of the *imago Dei*. He writes, "Theists…will say that the intelligibility of the universe is grounded in the nature of the ultimate rationality of God: both the universe and the human mind. It is, therefore, not surprising when the mathematical theories spun by human minds created in the image of God's Mind, find ready application in a universe whose architect was that same creative Mind."[27] Lennox regards the scientific enterprise itself as solid evidence for theism:

> One of the most powerful evidences, to my mind, that there is an eternal mind behind the universe is first of all that we can do science, that we can do it in the language of mathematics, that we have language we can use. We can use abstract concepts that are not material to describe things that are physical. All of that points in one direction, and one direction only, and it's this: 'In the beginning was the Word,' not the particles.[28]

These remarks are quite reminiscent of arguments made by scholars of earlier decades, such as the late theologian Thomas Torrance, who wrote extensively on the compatibility of Christianity and the sciences as well as the comprehensibility of the cosmos. "It is in and through the universe of space and time," Torrance writes,

---

[27] John Lennox, *God's Undertaker: Has Science Buried God?* (Oxford: Lion Books, 2009), 62.

[28] John Lennox, "The Question of Science and God—Part 1," Socrates in the City with Eric Metaxas, January 12, 2018.

"that God has revealed himself to us in modes of rationality that he has conferred upon the creation *and upon us in the creation*...This is God's universe, which he made accessible to our inquiries."[29] It is, then, God's act of bestowing upon mankind a type of rationality that is commensurate with the order of nature that enables us to perceive that order. Torrance continues:

> He created the universe and endowed man with gifts of mind and understanding to investigate and interpret it. Just as he made life to reproduce itself, so he has made the universe—with man as an essential constituent of it—in such a way that it can bring forth and articulate knowledge of itself...the pursuit of natural science is one of the ways in which man, the child of God, fulfills his distinctive function in the creation...Man as scientist can be spoken of as the priest of creation, whose office it is to interpret the books of nature written by the finger of God, to unravel the universe in its marvelous patterns and symmetries, and to bring it all into orderly articulation in such a way that it fulfills its proper end as the vast theater of glory in which the Creator is worshipped and hymned and praised by his creatures.[30]

Torrance echoes Kepler and many others in his description of the scientific man as a type of priest who exegetes God's creation. Mankind is "reason incarnate," and the mathematically describable, "harmoniously rational" universe (which includes man as the knower of the pervasive rational order) makes the scientific endeavor possible.[31]

The grand success of the natural sciences powerfully affirms Kepler's belief that the image-bearing incorporeal soul is what

---

[29] Thomas Torrance, *The Ground and Grammar of Theology* (New York: T&T Clark, 1980), 1. Emphasis mine.

[30] Ibid., 5–6.

[31] Thomas Torrance, *Christian Theology and Scientific Culture* (Eugene, OR: Wipf and Stock, 1980), 1, 31.

equips the natural philosopher for the investigation of nature—the ability to think God's thoughts after Him.

Conclusion

Johannes Kepler was the culmination of the Pythagorean-Platonic tradition and the pivotal figure in the rise of modern astronomy. While he was not the only natural philosopher of his time who sought to defend a Copernican model of the known universe, he was the first one to develop a true *physica coelestis* and to search for a universal law to account for celestial mechanics. Although some of his quintessentially Pythagorean-Platonic notions were incorrect, they were the Virgilian guides to his momentous mathematical epiphanies. All of his thinking was framed by his theological conviction that the higher calling of the natural philosopher is to elucidate the archetypes that pre-existed in the mind of God and are made manifest in the creation. Max Caspar repeatedly notes that it was this idea that "seized [Kepler] most vehemently and which influenced him throughout his whole life."[1] Kepler understood the profound theistic implications of a mathematically rational universe and mankind's ability to calcluate aspects of its structure and motions.

This book has had two main objectives. Part I demonstrated that Keplerian natural theology—the tripartite harmony of archetype, copy, and image that explains the observable resonance of mathematics, nature, and mind—has an illustrious intellectual pedigree. Some of the relevant themes, such as the mathematical

---

1 Max Caspar, *Kepler*, 377.

ordering of the cosmos, emerged within early Pythagoreanism and were grafted into Platonic thought. What we now refer to as the Pythagorean-Platonic tradition can be traced forward through a lineage of great thinkers, both pagans and Christians, who had a significant influence upon Kepler, largely through the curriculum and professors at Tübingen. Kepler's major works, including the *Mysterium Cosmographicum*, *Astronomia Nova*, and *Harmonice Mundi*, beautifully illustrate his intellectual debt to the great Western Tradition, and it is undeniable that Pythagorean-Platonic philosophy was crucial in the formation of his natural theology.

Part II made the case that today, Keplerian natural theology offers a more robust explanatory paradigm than ever before. It makes sense of what is otherwise an inexplicable yet powerful interconnection between disparate aspects of reality: abstract mathematical truths, the material realm to which those truths apply with startling precision, and the human rationality necessary for wielding mathematical tools in the natural sciences. Without this fortunate resonance, the natural sciences would have been impossible. A survey of some of the more prominent figures of the twentieth-century physics revolution revealed the persistence of philosophical questions surrounding the comprehensibility of the cosmos, and analyses of the contemporary conversations on God and abstract objects, the applicability of mathematics to the natural sciences, and philosophy of mind established the ongoing relevance and unprecedented potency of Keplerian natural theology.

Naturalism, as some non-theists have themselves acknowledged, must treat the deep connections between mathematics, nature, and mind as merely a fortunate convergence of accidental circumstances or deep, unsolvable mysteries. By contrast, Keplerian natural theology offers a philosophically rigorous and theologically sound explanation for this grand cosmic resonance. The tripartite harmony between the mind of God, His material creation, and the mind of man that inspired Kepler's life's work offers an intellectually satisfying metaphysical unity and conceptual coherence that naturalism does not, making it tremendously useful in the cumulative case for cosmic design. For these reasons, it should be regarded as one of the enduring treasures of Western thought. It is the liturgical refrain that adorns Kepler's priestly exposition of God's extraordinary book of nature.

Appendix

Kepler's Remaining Years and Legacy[1]

In October 1613, a little over a year after his arrival in Linz, Kepler married the twenty-four-year-old Susanna Reuttinger, who proved to be a good wife and a competent stepmother to his surviving children from his first marriage—eleven-year-old Susanna and six-year-old Ludwig.[2] Kepler worked diligently to classically educate his son, whose German-to-Latin translations of a fragment of Caesar's *Gallic War* and the first book of Tacitus's history still survive. Both children were thoroughly catechized in the Christian faith, and the memorization lesson Kepler composed about the Eucharist is preserved in the library at the University of Tübingen. By the spring of 1625, six more children had been born, but sadly, the first three died very young from various ailments.

In Linz, Kepler was banned from taking communion with the Lutheran congregation on account of his disagreement with the Formula of Concord. He was a deep and critical thinker about theological matters; he combed the works of great ancient theologians in his quest to determine correct doctrine on disputed points, and this led him to both agreements and disagreements with Lu-

---

1 With the exception of the footnoted passages, the content of this biographical summary has been distilled from Max Caspar's *Kepler*.

2 Regina, Kepler's stepdaughter from his first marriage, had married in 1608. Sadly, she died in 1617 at the age of twenty-seven.

theranism, Calvinism, and Catholicism. As a result, he was scorned by all of them, despite his constant insistence that his desire was for Christian unity, for a healing of the factions—something he prayed for daily. He believed that interpretive error should be weeded out no matter where it occurred and that it was egregious to exclude a brother in Christ from worship merely for disagreement with a secondary doctrine.

Despite his tremendous grief over these matters, Kepler settled into his work. He began outlining a comprehensive text on Copernican astronomy (more accurately, Keplerian astronomy) and published the first part of the *Rudolphine Tables*—which contained planetary tables and a star catalogue based upon Tycho Brahe's extensive observational records. Unfortunately, Kepler's scholarly momentum was hindered when, in 1615, his mother was accused of witchcraft. Using his status and connections to his advantage, Kepler took on the lengthy and arduous task of serving as Katharina's legal defense. The six-year ordeal, which required lengthy interruptions in his work, finally ended in his mother's favor, but she passed away less than a year after her release. During these already trying years, Kepler endured the loss of two daughters, the first a toddler and the second an infant. Compounding this heartache and hardship was the beginning of the terribly destructive Thirty Years' War, the end of which he would not live to see. Despite all of this, Kepler worked when he could, and managed to publish the first parts of his magisterial textbook, the *Epitome Astronomiae Copernicanae*, in 1618 and his most beloved brainchild, the *Harmonice Mundi*, in 1619.

In the spring of 1626, the city of Linz came under siege as the result of a violent peasant uprising, and the residents suffered from fires, hunger, and disease. A company of soldiers took up residence in the Kepler family's home, a situation that was disruptive and often harrowing: "The ears were constantly assailed by the noise of the cannon, the nose by evil fumes, the eye by flames. All doors had to be kept open for the soldiers, who, by their comings and goings, disturbed sleep at night, and work during the daytime."[3] Somehow, Kepler managed to find pockets of time to work, but publication ceased because of the destruction of the city's printing press, and he fretted over the preservation of his books and notes. The siege ended in August, but Linz was no longer a hospitable place for life and work. Fortunately, the emperor granted Kepler's request to relocate to Ulm, and he departed Linz with his family and all the worldly goods they could carry with them in November.

During the journey to Ulm, the boat came to an icy impasse, so Kepler installed his family in a residence in Regensburg and continued by wagon without them. Upon his arrival, he set to work, but experienced frustrating delays and disagreements with the printer he had commissioned for the completed *Rudolphine Tables*. To make matters worse, there was a conflict with the Brahe family over the title and dedication of the book that caused further complications and required a trip to Frankfurt. Finally, the first printing was completed in early September of 1627.

Kepler departed Ulm in late November and was reunited with his family, but then traveled to Prague only a month later to present the *Tables* to the emperor. Once at court, where he was received with honor, the emperor made him an attractive offer to stay—under the condition of conversion to Catholicism. However, Kepler could not, in good conscience, ignore crucial theological disagreements to gain the comfortable situation he had long yearned for. After a few months of navigating court politics, he was finally granted permission to settle in Sagan. After a trip to Linz to negotiate his formal release from his post as district mathematician, he and his household arrived in Sagan in May of 1628. Unfortunately, Kepler never quite adjusted to the culture and was starved for local intellectual community. Moreover, warfare broke out in the city,

---

3 A letter to Paul Guldin quoted in Arthur Koestler, *The Watershed: A Biography of Johannes Kepler* (New York: Anchor Books, 1960), 239.

and in November, non-Catholic citizens were ordered to convert or leave. As a foreigner with a commissioned appointment, Kepler was allowed to remain, but he was ostracized by friends and acquaintances who feared the association. For the next couple of years, he buried himself in work, avid correspondence, and in the establishment of a printing press, which the city lacked. In April of 1630, Kepler's wife Susanna gave birth to their seventh child, about a month after the celebrated marriage of his eldest daughter, Susanna, to astronomer Jakob Bartsch. Both events brought great joy to an otherwise bleak period of his personal life.

In October of 1630, amidst ongoing confessional warfare and political upheaval, Kepler set out for Linz, hoping to collect income that was due him. On the fourteenth he made a stop in Leipzig, where he lodged with a professor by the name of Philipp Müller. On the thirty-first, he penned a letter in which he mentioned his intent to shortly depart Leipzig "for Regensburg and Linz and from there to the Duke, that is to Sagan, if God will."[4] He did so on horseback, in the fog and icy conditions of a German winter. On November 2nd he made it to Regensburg, where he took up lodging with an acquaintance, but fell gravely ill within a few days. Just before his death on November 15th, in response to a pastor's question, Kepler declared that Jesus Christ was his hope of salvation, refuge, and solace.

Two days after his final breath, Kepler was memorialized by a large gathering of admirers, acquaintances, and friends, and then buried in the cemetery of St. Peter's (a Protestant church outside the city walls). His grave was marked, by some friends, with a modest tombstone inscribed with his self-composed epitaph:

> Once I measured the skies,
> now I measure the earth's shadow.
> Of heavenly birth was the measuring mind,

---

4 Carola Baumgardt, *Johannes Kepler: Life and Letters*, 191.

in the shadow remains only the body.[5]

Less than two years later, the churchyard at St. Peter's was destroyed in the Thirty Years' War. The exact location of Kepler's remains was permanently obscured. Max Caspar laments:

> Tycho Brahe's grave is in the Tyn church in Prague, Galileo is buried in the venerable church of Sante Croce in Florence, Newton rests among the great dead in Westminster Abbey. Veneration for genius erected these worthy monuments. But no tombstone covers the place where the no less gifted Kepler was interred. It is as though the fate, which in life gave him no peace, continued to pursue him even after death.[6]

Those who knew him expressed heartfelt lamentations over Kepler's untimely death and the great loss to the sciences. On November 25th, one of his friends wrote to another, "I cannot communicate [the news] to you with dry eyes: our mutual—alas!—former friend Kepler, a star of the first order in the mathematical sky, has passed away and rose above the horizon of earthly life...O, what immeasurable loss have the sciences suffered by the passing away of this incomparable man!"[7] The following January, Kepler's son-in-law, Jakob Bartsch, wrote to Philipp Müller: "I can hardly think of it without mournful tears in my eyes...O, woe! In the greatest disorder of his circumstances! O, woe! Your, mine, our sun, the sun of all astronomers has set, and has left to his people the darkness of sorrows, of struggle and confusion."[8]

Consistent with the theme of tragic circumstances that plagued Kepler during his life, it was not until early December that

---

5 James Voelkel, *Johannes Kepler and the New Astronomy* (New York: Oxford University Press, 1999), 130. Also see Endnote 17.

6 Max Caspar, *Kepler*, 361.

7 Baumgardt, *Johannes Kepler: Life and Letters*, 194.

8 Ibid., 195.

his family learned of his passing. Suddenly, his wife Susanna was a young widow with several children to care for, one still in infancy. Even the purchase of the traditional mourning clothes was a painful expense. Fortunately, Jakob Bartsch stepped into the role of family advocate and steward of Kepler's work, including the remaining unpublished manuscripts. He and his stepmother-in-law fought together to collect what was owed to the Kepler estate, but with limited success. During the associated travels, the family was finally able to visit Kepler's grave, nearly a year after his death.

Bartsch edited one of Kepler's surviving manuscripts, a science fiction story the latter had begun working on in 1608. Tragically, Bartsch died from the plague in 1633, before he could see to the publication of the work. This was another significant financial blow to the family, which became quite impoverished. Kepler's son Ludwig managed to finish preparing his father's manuscript and had it published in 1634 under the title *Somnium* (Latin, "The Dream"). The remainder of Kepler's scholarly legacy, containing thousands of fragments and sketches, astronomical calculations, and an enormous body of professional and personal correspondence, was held by Ludwig, who failed to publish any of it. Two of the younger Kepler children died in 1635, presumably from the plague, and their mother followed them in death in 1638 at the age of forty-seven, in the same city where her husband had died.

Miraculously, the Keplerania was preserved through a sale to a private individual and the subsequent fiery destruction of that scholar's home library. However, the manuscripts fell into obscurity for a time prior to their rediscovery in 1765. After several years of debate about the value of the collection, it was purchased with jewels by a Russian empress (and German princess), Catherine II, in 1773. She presented the collection to the Russian Academy of Science, which eventually turned it over to the Pulkova observatory in St. Petersburg (which was directed by a German). The Keplerian manuscript corpus now rests safely in the archives at Leningrad.

A monument to Kepler was erected in Regensburg in 1808, in the general vicinity of his original burial place. It was dedicated on December 27th, in celebration of the day of his birth.[9] He is now hailed as one of the mighty giants of the scientific revolution.

*The Kepler Monument in Regensburg, Germany.
Arcihtect: Emanuel Herigoyen.*

---

9   According to the Julian calendar.

Endnotes

1. The twentieth century brought important progress in the ongoing discussion about pre-Platonic Pythagoreanism. This was initiated by the publication of Walter Burkert's epoch-making treatise, *Lore and Science in Ancient Pythagoreanism*, which appeared in English translation in 1972. There has since been significant (though not voluminous) work done, notably by Carl Huffman and W. K. C. Guthrie, but there is by no means unanimous agreement with all of Burkert's conclusions, or even with his general characterization of Pythagoras the philosopher. Guthrie takes a more liberal view of what can reasonably be attributed to Pythagoras, but his reasoning is compelling. He writes, "The religious doctrines of immortality and transmigration are assigned to Pythagoras on incontrovertible positive evidence. His character as one of the most original thinkers in history, a founder of mathematical science and philosophical cosmology, although not directly attested by such early and impregnable sources, must be assumed as the only reasonable explanation of the unique impression made by his name on subsequent thought. It was both as a religious teacher and as scientific genius that he was from his own lifetime and for many centuries afterwards venerated by his followers, violently attacked by others, but ignored by none" (Guthrie, 181).

2. Walter Burkert has argued that certain fragments (1–7, 13, and 17) should be taken as authentic largely because the text is either consonant with the secondary information about Pythagoreanism found in Plato or Aristotle, or because it can only be understood in the context of pre-Socratic thought. See Burkert, 276–277 and Guthrie, 333. Note that J. E. Raven and G. S. Kirk in *The Presocratic Philosophers*, an authoritative work which predates Burkert, Guthrie, and Huffman, dismiss the Philonic fragments as "part of a post-Aristotelian forgery" (Graham, 46). This view was dominant in the mid-twentieth century when that text was published. The work of Burkert, Guthrie, and Huffman is more recent, and the fragments deemed reliable by Burkert (perhaps the most conservative of the three) include some that are of particular value to the present discussion.

3. Philolaus's cosmogony is elucidated in Fragment 6:

> Concerning nature and harmony the situation is this: the being of things, which is eternal, and nature itself admit of divine and not human knowledge, except that it was impossible for any of the things that are and are known by us to have come to be, if the being of the things from which the world-order came together, both the limiting things and the unlimited things, did not preexist. But since these beginnings preexisted and were neither alike nor even related, it would have been impossible for them to be ordered, if a harmony had not come upon them, in whatever way it came to be. Like things and related things did not in addition require any harmony, but things that are unlike and not even related, it is necessary that such things be bonded together by harmony, if they are going to be held in an order. (Graham, 50–51)

There is no scholarly consensus on what Philolaus means by "limiting things" and "unlimited things," what he views as the first principles or "beginnings" (*archai*) of things; Burkert suggests that these are akin to the atomic theory (atoms and the void) (Burkert, 259). In support of this view he cites a report that Democritus studied with a Pythagorean, which, if true, throws Philolaus and Leucippus into proximity (Burkert, 259). Huffman, however, sees the unlimited things as a kind of continuum and the limiting things as those which provide boundaries (Huffman, Philolaus of Croton, 43–44). For Philolaus, the unlimited and the limiters are essential ingredients of the cosmos, without which the world could not be known to human beings. Thus, they function in both an ontological and epistemological capacity. In a view that seems consonant with Huffman, Graham explains that Philolaus distinguishes between formal and material components:

> Philolaus here uses the term *archa* to mean something like "principle" or (theoretical) "starting-point." Thus he seems to attain a level of abstraction that surpasses that of many of his contemporaries and probably all of his predecessors. He does not yet recognize absolute starting-points of ontology (e.g. a category) or of explanation (an axiom), but he does recognize plural relative starting-points for the construction of the world and for explanations of its workings. (Graham, 62)

4. The idea of number being the "substance of all things" seems consistent with Fragment 4, but it may in fact go somewhat beyond Philolaic doctrine, depending upon the precise meaning of the phrase. Graham has suggested that Aristotle's reading here is incorrect:

> Much ink has been spilled on how "all is number" for the Pythagoreans. But Philolaus never makes this strong identity claim, and Aristotle's statements to this effect seem to arise from a misreading of the weaker claim embodied in fr. 4. Philolaus's princi-

ples are limiters and unlimited. Numbers probably belong to the class of limiters but they are not themselves principles of all things. (Graham, 54)

5. It is possible that this is not a precise understanding of Plato. Guthrie explains that

> the fact that Aristotle was able to equate Pythagorean mimesis with Plato's notion of physical objects as 'sharing in' the Ideas…should put us on our guard against the simple translation 'imitation'. The fact is, of course, that even Plato, and still more the Pythagoreans, were struggling to express new and difficult conceptions within the compass of an inadequate language (Guthrie, 230).

6. Plato has a thoroughly (but not exclusively) geometrical conception of the fundamental elements, in contrast with the broadly arithmetical view of Philolaus. Worthy of note, however, is the fact that Philolaic Fragment 12, which Burkert does not consider authentic, is taken by Guthrie as at least somewhat reflective of genuine Pythagorean thought. The fragment (preserved in Aëtius) reads: "The bodies in the sphere are five: fire, water, earth, and air, and fifthly the hull (?) of the sphere" (Guthrie, 267). Here we see the polyhedral theory of elements, including the fifth polyhedron as the outer boundary of the cosmos, found in the *Timaeus*. The phrase "bodies in the sphere" likely refers to the fact that the regular solids can be circumscribed by spheres. (Greek mathematician Theaetetus (c. 417–369 BC) is believed by some to have discovered methods for inscribing the regular solids within spheres. Euclid is thought to have drawn from Theaetetus's work in his own elucidation of the regular solids in Book XII of the *Elements*.) Guthrie argues that most likely, "Plato was here, as in so much else, adopting and elaborating Pythagorean notions" (Guthrie, 268). Burkert, however, points to the fact that

Aristotle was unaware of the polyhedral theory existing in Pythagorean thought (Burkert, 70). This is in reference to the *Metaphysics*, where Aristotle says that the Pythagoreans "have said nothing whatever about fire or earth or the other bodies of this sort" (Aristotle, *Metaphysics*, I.8).

7. Charlotte Methuen explains Melanchthon's philosophy of education:

> Melanchthon emphasises all seven liberal arts in the context of his educational programme...[and] he places an emphasis upon the place of the mathematical sciences in the curriculum which is unusual for educators of his time...his confidence in their utility is not his prime motive for including them in the educational curriculum. Far more important to him is the contribution made by mathematics, and particularly arithmetic and geometry, to the training of the mind in logical thinking and thus to the study of philosophy as a whole...he concludes, 'the first understanding is of number' (Methuen, *Kepler's Tübingen*, 71–72).

8. Bonaventure writes that "the observer considers things in themselves and sees in them weight, number, and measure...Hence he sees in them their mode, species, and order, as well as substance, power, and activity. From all these considerations the observer can rise, as from a vestige, to the knowledge of the immense power, wisdom, and goodness of the Creator." The mention of weight, number, and measure is notable in light of the intellectual history presented in the previous chapters. See Bonaventure, *The Journey of the Mind to God* (Indianapolis: Hackett Publishing, 1993), 8.

9. Maestlin did not teach Copernicanism in his regular classes at Tübingen. Rather, he reserved that controversial discussion for a select group of particularly advanced students,

which included Kepler. The official position of the university was geocentrism. Some of Maestlin's early biographers report that he experienced pushback from the theological authorities at Tübingen as a result of his views and teaching on Copernicanism. For a detailed discussion on the extent of Maestlin's inclusion of Copernicanism in his formal teaching, see Charlotte Methuen, "Maestlin's Teaching of Copernicus," *Isis* 87, no. 2 (June 1996): 230.

10. For the sake of extra income, Kepler wrote astrological calendars and cast horoscopes for his employers, even though he did not believe the celestial dynamics could actually be used to predict the future. He even made arguments for stripping astrology of what he called its "superstition." He did, however, believe that God had created the heavenly bodies to somehow have an influence on human life and that this could be explored empirically. See discussion in Caspar, 181–185. It is worthy of note that this idea of planetary influence on humanity is also found in Dante's *Purgatorio*.

11. In one scholarly source, *The Music of the Heavens: Kepler's Harmonic Astronomy* (Bruce Stephenson, 1994), the date of discovery is incorrectly stated as May 15, 1619 (129). From a careful reading of the statement in context, where the author mentions the actual publication of the *Harmonice*, it becomes clear that this erroneous date is an inadvertent typo rather than a misunderstanding, especially considering the fact that the author uses the correct date on an earlier page (125). Unfortunately, this error bled into another scholarly work, *Kepler's Philosophy and the New Astronomy* (Rhonda Martens, 2000), where the author (while citing Stephenson) deepens the error by stating that Kepler "finished the Harmonice on May 27th, 1618, but discovered the third law almost a year later (May 15, 1619). To include this important finding in the published version, Kepler

very quickly had to revise book V, with the typesetting already well underway" (Martens, 113). It is indeed true that Kepler spent months after his completion of the *Harmonice* doing revisions to the chapter containing the Third Law while the typesetting was being done for the preceding portion of the book. Typesetting was officially completed in February 1619. This means that Martens is incorrect about the discovery occurring in May 1619.

12. It is interesting to note (though the translators of the *Harmonice* do not) that Kepler's Egyptian treasure analogy is not original to him; it echoes several prominent Church Fathers who regarded pagan philosophy as a treasure trove of ideas that are useful in explicating and supporting Christian theology. For example, in *A Letter From Origen to Gregory*, Origen of Alexandria (184–253), a Christian Platonist, writes:

> I wish to ask you to extract from the philosophy of the Greeks what may serve as a course of study or a preparation for Christianity, and from geometry and astronomy what will serve to explain the sacred Scriptures, in order that all that the sons of the philosophers are wont to say about geometry and music, grammar, rhetoric, and astronomy, as fellow-helpers to philosophy, we may say about philosophy itself, in relation to Christianity. Perhaps something of this kind is shadowed forth in what is written in Exodus from the mouth of God, that the children of Israel were commanded to ask from their neighbours, and those who dwelt with them, vessels of silver and gold, and raiment, in order that, by spoiling the Egyptians, they might have material for the preparation of the things which pertained to the service of God. [Origen, *A Letter From Origen to Gregory*, in Ante-Nicene Fathers 4 (Peabody, MA: Hendrickson Publishers, 2012), 393.]

As discussed in Chapter 3, it is known with certainty that Kepler studied the Church Fathers, and he explicitly named Origen among them. It seems probable that Augustine gleaned this analogy from Origen. Caspar (61, 216) mentions the "Fathers of the Church" in relation to the deep theological studies Kepler carried out while trying to discern the truth about heavily disputed doctrinal points but does not specify *where* Kepler names Origen. It is plausible that it was in a letter Kepler wrote to his former teacher, Michael Maestlin, on December 22, 1616. A translated portion of this letter is found in Baumgardt (107), and there Kepler writes about his inability to confirm certain doctrines of the Council of Trent, saying that they "cannot be found in the old Fathers of the Church." However, in the same context, Caspar offers a short quote from a letter Kepler wrote to another former teacher, Matthias Hafenreffer, on April 11, 1619. Only a small portion of this letter is translated in Baumgardt, and it does not contain any mention of Origen or the Church Fathers in general.

13. This remarkable quote, which reflects Newton's philosophical response to the limitations of man's knowledge and the immensity of creation, inspired the work of the American Pulitzer Prize-winning poet, Richard Wilbur, who wrote in his poem "Worlds": "But Newton, who had grasped all space, was more/Serene. To him it seemed that he'd but played/With a few shells and pebbles on the shore/Of that profundity he had not made." Richard Wilbur, *Collected Poems: 1943–2004* (New York: Harvest Publishers, 2004), 111.

14. Kepler specifically sought natural causes as opposed to the ancient Greek idea of a pervasive, animating Word Soul. In a letter to Johann Georg Brengger dated October 4, 1607, Kepler wrote, "I furnish a heavenly philosophy (or physics) in place of the heavenly theology or metaphysics of Aristotle." See Carola Baumgardt, *Johannes Kepler: Life*

and Letters (New York: Philosophical Library, Inc., 1951), 75. This should not be confused with the concept of the "earth soul" which Kepler did affirm. See Aiton, Duncan, and Field's discussion in their introduction to *The Harmony of the World* (Philadelphia: American Philosophical Society, 1997), xxxii.

15. The clockwork metaphor became increasingly popular in the seventeenth century. The father of chemistry (and devout Christian) Robert Boyle (1627–1691) insisted that the world was not like a puppet that required the constant movement of a puppet master; it was like a grand clock in which the parts are so skillfully contrived that once they are set in motion by their maker, they proceed according to his design. See Boyle, *A Free Enquiry Into the Received Notion of Nature* (1682).

16. It is quite interesting that Kepler's argument anticipates a relatively recent one set forth by astronomer Guillermo Gonzalez and philosopher Jay Richards in *The Privileged Planet*. Their thesis is that "the cosmos, our Solar System, and our exceptional planet are themselves a laboratory, and Earth is the best bench in the lab." They devote a chapter to how earth's position in the solar system is the best for both survival of intelligent life and astronomy. See Guillermo Gonzalez and Jay Richards, *The Privileged Planet* (Washington, D.C.: Regnery Publishing, Inc., 2004), xv.

17. Kepler's self-written epitaph reflects his conviction that mind, with its mathematical abilities, bears the image of the divine and, upon death, departs from the earth-bound body: "Once I measured the skies, now I measure the earth's shadow. Of heavenly birth was the measuring mind, in the shadow remains only the body." James Voelkel, *Johannes Kepler and the New Astronomy* (New York: Oxford University Press, 1999), 130. Kepler's grave site was destroyed during the Thirty Years' War, but a close friend

made a surviving sketch of the tombstone. This sketch is included in Voelkel's book.

Sketch of Kepler's Tombstone

18. Einstein often made enigmatic statements about his beliefs regarding the existence and nature of God, and these have been interpreted in different ways. Einstein was dissatisfied with an atheistic or pantheistic characterization of his religious views; "I'm not an atheist and I don't think I can call myself a pantheist," he said, even though he expressed fascination with "Spinoza's God." He said, "I believe in Spinoza's God who reveals himself in the orderly harmony of what exists, not in a god who concerns himself with fates and actions of human beings." See Jammer, 48–49.

19. Eddington writes, "Since I cannot avoid introducing this question of a beginning, it has seemed to me that the most satisfactory theory would be one which made the beginning *not too unaesthetically abrupt*...the primordial state of

things which I picture is an even distribution of protons and electrons, extremely diffuse and filling all (spherical) space, remaining nearly balanced for an exceedingly long time until its inherent instability prevails." At the end of the book he admits, "The beginning seems to present insuperable difficulties unless we agree to look on it as frankly supernatural. We may have to let it go at that" (*Expanding Universe*, 125). However, Eddington still saw God as the ground of an eternal universe's existence.

20. Wheeler did not actually invent the term "black hole." In 1967, while giving a lecture at NASA's Goddard Institute astrophysics conference in New York, an audience member suggested it. A few weeks later, on December 29, 1967, Wheeler gave a lecture for the annual meeting of the American Association for the Advancement of Science in New York in which he used the term "black hole." The written version of the lecture was published the following spring in *American Scientist* 56, no. 1 (1968) and the term was thereby introduced into the scientific literature. See Carlos Herdeiro and José Lemos, "The black hole fifty years after: Genesis of the name," in *History and Philosophy of Physics*, December 12, 2019.

21. Haldane went on to say, "But as regards my own very finite and imperfect mind, I can see, by studying the effects on it of drugs, alcohol, disease, and so on, that its limitations are largely at least due to my body. Without that body it may perish altogether, but it seems to me quite as probable that it will lose its limitations and be merged into an infinite mind or something analogous to a mind which I have reason to suspect probably exists behind nature. How this might be accomplished I have no idea" (209). He then reaffirms his agnosticism about the reality of an afterlife.

22. The young philosopher Elizabeth Anscombe famously leveled a major critique at the first version of Lewis's AR,

and he subsequently made some key revisions. Anscombe's original critique and a brief postscript response from Lewis are recorded in her essay, "A Reply to Mr. C. S. Lewis's Argument that 'Naturalism' is Self-Refuting," in Metaphysics and the Philosophy of Mind: Collected Philosophical Papers Vol. 2. (Minneapolis: University of Minnesota Press, 1981).

Bibliography

Adler, Mortimer. "Biographical Note: Nicomachus." *Great Books of the Western World 10*. Chicago: Encyclopaedia Britannica, Inc., 1990.

———. "Biographical Note: Saint Augustine." *Great Books of the Western World 16*. Chicago: Encyclopaedia Britannica, Inc., 1990.

———. "Biographical Note: Sir Arthur Eddington." *Great Books of the Western World 56*. Chicago: Encyclopaedia Britannica, Inc., 1990.

———. *The Great Ideas: From the Great Books of Western Civilization*. Chicago: Open Court, 2001.

Albertson, David. *Mathematical Theologies: Nicholas of Cusa and the Legacy of Thierry of Chartres*. Oxford: Oxford University Press, 2014.

Allison, Henry E. "Kant's Refutation of Materialism." *The Monist* 72, 2 (April 1, 1989): 190–208.

Alston, William P. *Perceiving God: The Epistemology of Religious Experience*. Ithaca, NY: Cornell University Press, 1991.

Anscombe, Elizabeth. "A Reply to Mr. C. S. Lewis's Argument that 'Naturalism' is Self-Refuting." In *Metaphysics and the Philosophy of Mind: Collected Philosophical Papers Vol. 2*. Minneapolis: University of Minnesota Press, 1981.

Aristotle. *Metaphysics*. Great Books of the Western World 7. Chicago: Encyclopaedia Britannica, 1990.

———. *On the Heavens*. Great Books of the Western World 7. Chicago: Encyclopaedia Britannica, 1990.

Athanasius. *On the Incarnation* with *Against the Heathen*. Brookline, MA: Paterikon Publications, 2018.

Atkins, Peter. *Creation Revisited*. London: Penguin Books, 1994.

Augustine. *The City of God*. Great Books of the Western World 16. Chicago: Encyclopaedia Britannica, Inc., 1990.

———. *The Confessions*. Great Books of the Western World 16 (Chicago: Encyclopaedia Britannica, Inc., 1990.

———. *Eighty-three Different Questions*. In *The Fathers of the Church* 70. Washington, DC: CUA Press, 2010.

———. *On Christian Doctrine*. Great Books of the Western World 16. Chicago: Encyclopaedia Britannica, Inc., 1990.

———. *On Christian Teaching*. New York: Oxford University Press, 2008.

———. *On Free Choice of the Will*. Indianapolis, IN: Hackett Publishing Co., 1993.

———. *On Genesis*. New York: New City Press, 2002.

———. *Sermons: 51–94*. Hyde Park, NY: New City Press, 1991.

Barker, Andrew. "Mathematical Beauty Made Audible: Musical Aesthetics in Ptolemy's *Harmonics*." *Classical Philology* 110 (October 2010): 403–420.

———. "Ptolemy's Pythagoreans, Archytas, and Plato's conception of mathematics." *Phronesis* 39, no. 2 (1994): 113–135.

Balfour, Arthur James. *The Foundations of Belief.* New York: Longmans, Green, & Co., 1895.

Barker, Peter and Bernard R. Goldstein. "Theological Foundations of Kepler's Astronomy." *Osiris* 16 (2001): 88–113.

Barlow, Nora, ed. *The Autobiography of Charles Darwin 1809–1882*. London: W. W. Norton & Company, 1958.

Baumgardt, Carola. *Johannes Kepler: Life and Letters*. New York: Philosophical Library, 1951.

Berman, Sophie. "Human Free Will in Anselm and Descartes." *The Saint Anselm Journal* 2.1 (Fall 2004): 1-9.

Blasi, Anthony, Jean Duhaime, and Paul-Andre Turcotte. *Handbook of Early Christianity: Social Science Approaches*. Walnut Creek, CA: AltaMira Press, 2002.

Boethius. *Consolation of Philosophy*. New York: Oxford University Press, 2008.

Bonaventure. *The Journey of the Mind to God*. Indianapolis: Hackett Publishing, 1993.

Boner, Patrick J. "Life in the Liquid Fields: Kepler, Tycho and Gilbert on the Nature of the Heavens and Earth." *History of Science* 46, no. 3 (2008): 275–297.

Boyle, Robert. *A Free Enquiry Into the Received Notion of Nature* (1682). Accessed December 20, 2018 at http://downloads.it.ox.ac.uk/ota-public/tcp/Texts-HTML/free/A28/A28982.html.

Brian, Denis. *Einstein: A Life*. New York: John Wiley & Sons, 1996.

Broadie, Sarah. *Nature and Divinity in Plato's Timaeus*. Cambridge: Cambridge University Press, 2011.

Burkert, Walter. *Lore and Science in Ancient Pythagoreanism*. Cambridge: Harvard University Press, 1972.

Bussey, Peter. Interview in *God and the Big Bang: how the universe began—the moment of creation*. Focus Media. Accessed February 3, 2019 at https://www.youtube.com/watch?v=nBulsNbaYgo.

———. *Signposts to God: How Modern Physics and Astronomy Point the Way to Belief*. Downers Grove: IVP Academic, 2016.

Caspar, Max. *Kepler*. New York: Dover Publications, 1993.

Chesterton, Gilbert Keith. *The Well and the Shallows*. In *The Collected Works of G. K. Chesterton*, edited by James J. Thompson. San Francisco: Ignatius Press, 1990.

Cho, Adrian. "The Discovery of the Higgs Boson." *Science* 338 (December 21, 2012): 1524–1525.

Churchland, Paul. *Matter and Consciousness*. Cambridge: MIT Press, 2013.

Classens, Guy. "Imagination as Self-knowledge: Kepler on Proclus' Commentary on the First Book of Euclid's Elements." *Early Science and Medicine* 16 (2011): 179–199.

Clement of Alexandria. *The Stromata*. Ante-Nicene Fathers 2. Edited by Alexander Roberts and James Donaldson. Peabody, MA: Hendrickson Publishers, 2012.

Cochrane, Charles. *Christianity and Classical Culture*. Indianapolis: Oxford University Press, 1940.

Colyvan, Mark. "The Miracle of Applied Mathematics." *Synthese* 127, no. 3 (2001): 265–277.

Copan, Paul and William Lane Craig. *Creation Out of Nothing: A Biblical, Philosophical, and Scientific Exploration.* Grand Rapids: Baker Academic, 2004.

Copernicus, Nicolaus. *Revolutions of the Heavenly Spheres.* Great Books of the Western World, 15. Chicago: Encyclopaedia Britannica, Inc., 1990.

Cox, Ronald. *By the Same Word: Creation and Salvation in Hellenistic Judaism and Early Christianity.* Berlin: Walter De Gruyter, 2007.

Coyne, George and Michael Heller. *A Comprehensible Universe: The Interplay of Science and Theology.* New York: Springer, 2010.

Craig, William Lane. "Absolute Creationism and Divine Conceptualism: A Call for Conceptual Clarity." *Philosophia Christi* 19, no. 2 (2017): 431–438.

———. *God Over All.* New York: Oxford University Press, 2016.

———. "Nominalism and Divine Aseity." In *Oxford Studies in Philosophy of Religion* 4. Edited by Jonathan L. Kvanvig. New York: Oxford University Press, 2012.

———. 2016. "Roger Penrose Interview, Part 1." May 8, 2016. Transcript of *Reasonable Faith* podcast, accessed February 8, 2019 at https://www.reasonablefaith.org/media/reasonable-faith-podcast/roger-penrose-interview-part-1/.

Criddle, A. H. "The Chronology of Nicomachus of Gerasa." *The Classical Quarterly* 48, no. 1 (1998): 324–327.

Danielson, Dennis. "The great Copernican cliche." *American Journal of Physics* 69, no. 10 (October 2001): 1029–1035.

Darwin, Charles. "To William Graham: 3 July 1881." Darwin Correspondence Project at Cambridge University, accessed February 19, 2019 at https://www.darwinproject.ac.uk/letter/DCP-LETT-13230.xml.

Davies, Paul. *Are We Alone?* New York: Orion Productions, 1995.

———. *The Goldilocks Enigma: Why is the Universe Just Right for Life?* Boston: Houghton Mifflin, 2008.

Dennett, Daniel. *From Bacteria to Bach and Back: The Evolution of Minds.* New York: W.W. Norton & Company, 2017.

Descartes, René. *Meditations.* Great Books of the Western World 28. Chicago: Encyclopaedia Britannica, Inc., 1990.

Dillon, John. *The Middle Platonists: 80 BC to AD 220.* Ithaca: Cornell University Press, 1996.

Dirac, Paul. "The Evolution of the Physicists' Picture of Nature." *Scientific American* 208, no. 5 (May 1963): 45–53.

Dretske, Fred. "Reasons and Causes." *Philosophical Perspectives* 3 (1989): 1–15.

Dreyer, J. L. E. *A History of Astronomy from Thales to Kepler.* New York: Dover, 1953.

Eddington, Arthur. *The Expanding Universe.* New York: Cambridge University Press, 1988.

———. *The Nature of the Physical World.* New York: Cambridge University Press, 1958.

———. *Science and the Unseen World.* New York: Macmillan Company, 1929.

Bibliography

Einstein, Albert. *Letters to Maurice Solovine.* Edited by Neil Berger. Paris: Gauthier-Villars, 1956.

———. "Physics and Reality." *Daedalus* 132, no. 4 (Fall, 2003): 22–25.

———. *The World as I See It.* New York: Kensington, 2006.

Ferguson, Kitty. *Pythagoras: His Lives and the Legacy of a Rational Universe.* London: Icon Books, 2011.

Ferngren, Gary, ed. *Science & Religion: A Historical Introduction.* Baltimore: Johns Hopkins University Press, 2017.

Field, J. V. *Kepler's Geometrical Cosmology.* New York: Bloomsbury, 1988.

Finocchiaro, Maurice A., editor and translator. *The Essential Galileo.* Indianapolis: Hackett Publishing, 2008.

Galilei, Galileo. *Dialogue Concerning the Two Chief World Systems.* Translated by Stillman Drake. Berkeley: University of California, 1962.

Gingerich, Owen. *The Book Nobody Read: Chasing the Revolutions of Nicolaus Copernicus.* New York: Walker, 2004.

———. *Copernicus.* New York: Oxford University Press, 2016.

———. "Creative Revolutionaries: How Galileo and Kepler Changed the Face of Science." *Euresis* 2 (Winter, 2012): 9–17.

———. *The Eye of Heaven: Ptolemy, Copernicus, Kepler.* New York: American Institute of Physics, 1993.

———. "Kepler and the Laws of Nature," *Perspectives on Science and Christian Faith* 63, no. 1 (March 2011): 17–23.

———. "Kepler Then and Now." *Perspectives on Science* 10, no. 3 (2002): 228–240.

Goetz, Stewart. "The Argument from Reason." *Philosophia Christi* 15, no. 1 (2013): 47–62.

Gonzalez, Guillermo and Jay Richards. *The Privileged Planet*. Washington, D.C.: Regnery Publishing, Inc., 2004.

Gordon, Bruce. "Eddington, Arthur." In *Dictionary of Christianity and Science*. Edited by Paul Copan et al. Grand Rapids: Zondervan, 2017.

Gould, Paul, ed. *Beyond the Control of God: Six Views on the Problem of God and Abstract Objects*. New York: Bloomsbury Academic, 2014.

———. (2010) "A Defense of Platonic Theism," Doctoral Dissertation, Purdue University, 2010, Database (AAI3413791).

———. "The Problem of God and Abstract Objects: A Prolegomenon." *Philosophia Christi* 13, no. 2 (2011): 255–274.

———. "Theistic Activism and the Doctrine of Creation." *Philosophia Christi* 16, no. 2 (2014): 283–296.

Graham, Daniel W. "On Philolaus's astronomy." *Archive for History of Exact Sciences* 69 (2015): 217–230.

———. "Philolaus," in *A History of Pythagoreanism*. Edited by Carl Huffman. Cambridge: Cambridge University Press, 2014.

Grattan-Guinness, Ivor. "Solving Wigner's Mystery: the Reasonable (Though Perhaps Limited) Effectiveness of Mathematics in the Natural Sciences." *The Mathematical Intelligencer* 30, no. 3 (2008): 7–17.

Greene, Brian. *The Elegant Universe: Superstrings, Hidden Dimensions, and the Quest for the Ultimate Theory*. New York: W. W. Norton & Co., 2003

Gregory Thaumaturgus. *Oration and Panegyric Addressed to Origen.* Ante-Nicene Fathers 6. Edited by Alexander Roberts and James Donaldson. Peabody, MA: Hendrickson Publishers, 2012.

Gregory, Andrew. "The Pythagoreans: Number and Numerology," in *Mathematicians & Their Gods.* Edited by Snezana Lawrence and Mark McCartney. Oxford: Oxford University Press, 2015.

Guthrie, W. K. C. *A History of Greek Philosophy: The Earlier Presocratics and the Pythagoreans.* Cambridge: Cambridge University Press, 1988.

Haldane, J. B. S. *The Inequality of Man.* London: Chatto & Windus, 1932.

———. *Possible Worlds.* New York: Routledge, 2017.

Hall, A. R. and M. B. Hall, ed. *Unpublished Scientific Papers of Isaac Newton.* Cambridge University Press, 1962.

Hannam, James. *The Genesis of Science: How the Christian Middle Ages Launched the Scientific Revolution.* Washington, DC: Regnery Publishing, 2011.

Hardy, G. H. *A Mathematician's Apology.* Cambridge: Cambridge University Press, 2012.

Herdeiro, Carlos and José Lemos. "The black hole fifty years after: Genesis of the name." In the History and Philosophy of Physics category. Accessed February 9, 2019 at https://arxiv.org/pdf/1811.06587.pdf.

Hicks, Andrew. "Pythagoras and Pythagoreanism in late antiquity and the Middle Ages," in Carl Huffman, *A History of Pythagoreanism.* Cambridge: Cambridge University Press, 2014.

Holton, Gerald. *Thematic Origins of Scientific Thought.* Cambridge: Harvard University Press, 1988.

Howell, Russell W. "The Matter of Mathematics." *Perspectives on Science and Christian Faith* 67 no. 2 (June 2015): 74–88.

Huffman, Carl, ed. *A History of Pythagoreanism*. Cambridge: Cambridge University Press, 2014.

———. *Archytas of Tarentum: Pythagorean, Philosopher, and Mathematician King*. New York, Cambridge University Press, 2005.

———. *Philolaus of Croton: Pythagorean and Presocratic*. New York: Cambridge University Press, 1993.

———. "The Role of Number in Philolaus' Philosophy." *Phronesis* 33, no. 1 (1988): 1–30.

Hut, Piet, Mark Alford, and Max Tegmark. "On Math, Matter and Mind." *Foundations of Physics* 36 no. 6 (June 2006): 765–794.

Irenaeus. *Against Heresies*. Ante-Nicene Fathers 1. Edited by Alexander Roberts and James Donaldson. Peabody, MA: Hendrickson Publishers, 2012.

Jammer, Max. *Einstein and Religion*. Princeton, NJ: Princeton University Press, 1999.

Johansen, T. K. *Plato's Natural Philosophy: A Study of the Timaeus-Critias*. Cambridge: Cambridge University Press, 2008.

John of Salisbury. *The Metalogicon*. Philadelphia: Paul Dry Books, 2009.

Kahn, Charles. *Pythagoras and the Pythagoreans: A Brief History*. Indianapolis: Hackett Publishing Company, Inc., 2001.

Kaiser, Christopher B. *Toward a Theology of Scientific Endeavour: The Descent of Science*. Burlington, VT: Ashgate Publishing, 2007.

Kepler, Johannes. *Astronomia Nova*. Translated by William Donahue. Santa Fe, NM: Green Lion Press, 2015.

———. *The Harmony of the World*. Edited by E. J. Aiton, A. M. Duncan, and J. V. Field. Philadelphia: American Philosophical Society, 1997.

———. *Kepler's Conversation with Galileo's Sidereal Messenger*. Translated by Edward Rosen. Johnson Reprint Corp., 1965.

———. *Mysterium Cosmographicum*. Norwalk, CT: Opal Publishing, 1981.

Kim, Jaegwon. "Emergence: Core ideas and issues." *Synthese* 151 (2006): 547–559.

Kline, Morris. *Mathematics and the Physical World*. New York: Dover Publications, 1959.

Koestler, Arthur. *The Watershed: A Biography of Johannes Kepler*. New York: Anchor Books, 1960.

Kuhn, Thomas. *The Copernican Revolution*. Cambridge: Harvard University Press, 1995.

Lakoff, George and Rafael Núñez. *Where Mathematics Comes From*. New York: Basic Books, 2000.

Lawrence, Snezana and Mark McCartney ed. *Mathematicians & Their Gods: Interactions Between Mathematics and Religious Beliefs*. Oxford: Oxford University Press, 2015.

Layne, Danielle, ed. *Proclus and His Legacy*. Boston: De Gruyter, 2017.

Leftow, Brian. *God and Necessity*. Oxford: Oxford University Press, 2010.

Leibniz, G. W. *Philosophical Essays*. Indianapolis: Hackett Publishing, 1989.

Lennox, John. *God's Undertaker: Has Science Buried God?* Oxford: Lion Books, 2009.

———. "The Question of Science and God—Part 1." Socrates in the City, January 12, 2018. Accessed February 12, 2019 at https://socratesinthecity.com/watch/john-lennox-the-question-of-science-and-god-part-1/

Lewis, C. S. *The Discarded Image*. New York: Cambridge University Press, 2012.

———. *Miracles*. New York: HarperCollins, 2015.

———. *The Weight of Glory*. New York: HarperOne, 2000.

Lindberg, David. *The Beginnings of Western Science*. Chicago: University of Chicago Press, 2007.

Lissauer, Jack J. "In Retrospect: Kepler's *Astronomia Nova*." *Nature* 462 (December 2009): 725.

Locke, John. *Essay Concerning Human Understanding*. Indianapolis: Hackett Publishing, 1996.

Maestlin, Michael. "Maestlin to Kepler, 21 September 1616." *Johannes Kepler Gesammelte Werke* 17. Edited by Max Caspar. Munich: Verlag, 1955.

Markos, Louis. *From Plato to Christ: How Platonic Thought Shaped the Christian Faith*. Downers Grove: IVP Academic, 2021.

Martens, Rhonda. *Kepler's Philosophy and the New Astronomy*. Princeton: Princeton University Press, 2000.

———. "Kepler's Solution to the Problem of a Realist Celestial Mechanics." *Studies in History and Philosophy of Science* 30, no. 3 (1999): 377–394.

Martyr, Justin. *Hortatory Address to the Greeks.* Ante-Nicene Fathers 1. Edited by Alexander Roberts and James Donaldson. Peabody, MA: Hendrickson Publishers, 2012.

McCann, Hugh. *Creation and the Sovereignty of God.* Indianapolis: Indiana University Press, 2012.

McGrath, Alister. *Christian Theology.* West Sussex: Blackwell Publishers, Ltd., 2017.

———. *Science & Religion: A New Introduction.* West Sussex: Blackwell Publishers, Ltd., 2010.

Menuge, Angus. "Knowledge of Abstracta: A challenge to Materialism." *Philosophia Christi* 18, no. 1 (2016): 7–27.

Menzel, Christopher. "Theism, Platonism, and the Metaphysics of Mathematics." *Faith and Philosophy* 4, no. 4 (1987): 365–382.

Methuen, Charlotte. *Kepler's Tübingen.* Brookfield, VT: Ashgate Publishing, 1998.

———. "Maestlin's Teaching of Copernicus." *Isis* 87, no. 2 (June 1996): 230–247.

———. "The Role of the Heavens in the Thought of Philip Melanchthon," *Journal of the History of Ideas* 57, no. 3 (July, 1996): 385–403.

Mohr, Richard D. "Plato's Theology Reconsidered: What the Demiurge Does." *History of Philosophy Quarterly* 2, no. 2 (April, 1985): 131–144.

Moreland, J. P. *Consciousness and the Existence of God.* New York: Routledge, 2009.

———. and William Lane Craig. *Philosophical Foundations for a Christian Worldview.* Downers Grove, IL: IVP Academic, 2003.

———. *The Recalcitrant Imago Dei*. London: SCM Press, 2009.

Morris, Thomas and Christopher Menzel. "Absolute Creation." *American Philosophical Quarterly* 23, no. 4 (October 1986): 353–362.

Nagel, Thomas. *Mind and Cosmos: Why the Materialist Neo-Darwinian Conception of Nature is Almost Certainly False*. New York: Oxford University Press, 2012.

Newton, Isaac. "Letter to Edmund Halley, June 20, 1686." Accessed December 13, 2018 at http://www.newtonproject.ox.ac.uk/view/texts/normalized/NATP00325.

———. General Scholium to the *Principia Mathematica*, 3rd ed., 1726. Accessed March 18, 2019 at https://isaac-newton.org/general-scholium/.

Nicholas of Cusa. *Complete Philosophical and Theological Treatises of Nicholas of Cusa: Volume One*. Translated and edited by Jasper Hopkins. Minneapolis: Banning Press, 1990.

Nicomachus of Gerasa. *Introduction to Arithmetic*. Great Books of the Western World 10. Chicago: Encyclopaedia Britannica, Inc., 1990.

"The Nobel Prize in Physics 1963." Accessed January 31, 2019 at https://www.nobelprize.org/prizes/physics/1963/summary/.

"The Nobel Prize in Physics 2013." Accessed February 3, 2019 at https://www.nobelprize.org/prizes/physics/2013/summary/.

O'Meara, John J. "The Neoplatonism of Saint Augustine." In *Neoplatonism and Christian Thought*, edited by Dominic J. O'Meara. Norfolk: International Society for Neoplatonic Studies, 1982.

Omodeo, Pietro Daniel. "The 'Impiety' of Kepler's shift from mathematical astronomy to celestial physics." *Annalen der Physik* 527, no. 7–8 (2015): A63–A80.

Origen. *A Letter From Origen to Gregory*. Ante-Nicene Fathers 4. Edited by Alexander Roberts and James Donaldson. Peabody, MA: Hendrickson Publishers, 2012.

Parrish, Stephen E. "Defending Theistic Conceptualism." *Philosophia Christi* 20, no. 1 (2018): 101–117.

Penrose, Roger. "Mathematics, the Mind, and the Physical World." In *Meaning in Mathematics*, edited by John Polkinghorne. New York: Oxford University Press, 2011.

———. *Shadows of the Mind: A Search for the Missing Science of Consciousness*. New York: Oxford University Press, 1994.

Pesic, Peter. *Music and the Making of Modern Science*. Cambridge: MIT Press, 2014.

Philo. *The Works of Philo: Complete and Unabridged*. Edited by C. D. Yonge. Peabody, MA: Hendrickson Publishers, 1993.

Planck, Max. *Scientific Autobiography and Other Papers*. Great Books of the Western World 56. Chicago: Encyclopaedia Britannica, Inc., 1990.

Plantinga, Alvin. *The Nature of Necessity*. New York: Oxford University Press, 1982.

———. *Where the Conflict Really Lies: Science, Religion, and Naturalism*. New York: Oxford University Press, 2011.

Plato. *Phaedo*. Great Books of the Western World 6. Chicago: Encyclopaedia Britannica, 1990.

———. *The Republic*. Great Books of the Western World 6. Chicago: Encyclopaedia Britannica, 1990.

———. *Timaeus*. Great Books of the Western World 6. Chicago: Encyclopaedia Britannica, 1990.

Plotinus. *The Enneads*. Great Books of the Western World 11. Chicago: Encyclopaedia Britannica, 1990.

Plutarch. *Moralia, Volume XIII: Part 1: Platonic Essays*. Translated by Harold Cherniss. Cambridge, MA: Harvard University Press, 1976.

Polkinghorne, John. *Science and Creation: The Search for Understanding*. Philadelphia: Templeton Foundation Press, 2006.

———. *Theology in the Context of Science*. New Haven: Yale University Press, 2009.

Proclus. *A Commentary on the first Book of Euclid's Elements*. Translated by Glenn R. Morrow. Princeton: Princeton University Press, 1970.

Ptolemy. *The Almagest*. Great Books of the Western World 15. Chicago: Encyclopaedia Britannica, 1990.

Reid, Thomas. *Essays on the Intellectual Powers of Man*. Cambridge: John Bartlett, 1852.

Reppert, Victor. *C. S. Lewis's Dangerous Idea: In Defense of the Argument from Reason*. Downers Grove: InterVarsity Press, 2003.

Rickabaugh, Brandon and Todd Buras. "The Argument from Reason, and Mental Causal Drainage: A Reply to Peter van Inwagen." *Philosophia Christi* 19, no. 2 (2017): 381–399.

Rothman, Aviva. *The Pursuit of Harmony: Kepler on Cosmos, Confession, and Community*. Chicago: University of Chicago Press, 2017.

Rublack, Ulinka. *The Astronomer & the Witch: Johannes Kepler's Fight for his Mother*. New York: Oxford University Press, 2015.

Runia, David. *On the Creation of the Cosmos According to Moses: Introduction, Translation and Commentary*. Atlanta: Society of Biblical Literature, 2001.

———. *Philo of Alexandria and the Timaeus of Plato*. Leiden, The Netherlands: E. J. Brill, 1986.

———. "Why Does Clement of Alexandria Call Philo 'The Pythagorean'?" *Vigiliae Christianae* 49, no. 1 (March, 1995): 1–22.

Russell, Bertrand. *Principles of Mathematics* I. London: Cambridge University Press, 1903.

Schenck, Kenneth. *A Brief Guide to Philo*. Louisville: John Knox Press, 2005.

Se'eman, Yuval. "Plato Alleges that God Forever Geometrizes," *Foundations of Physics* 26, no. 5 (1996): 575-583.

Spence, Joseph and John Underhill. *Spence's Anecdotes, Observations, and Characters of Books and Men*. London: W. Scott, 1890.

Stanley, Matthew. *Practical Mystic: Religion, Science, and A. S. Eddington*. Chicago: University of Chicago Press, 2007.

Steiner, Mark. *The Applicability of Mathematics as a Philosophical Problem*. Cambridge: Harvard University Press, 1998.

Stephenson, Bruce. *Kepler's Physical Astronomy*. Princeton, Princeton University Press, 1994.

———. *The Music of the Heavens: Kepler's Harmonic Astronomy*. Princeton: Princeton University Press, 1994.

Tanzella-Nitti, Giuseppe. "The Two Books Prior to the Scientific Revolution." *Perspectives on Science and Christian Faith* 57, no. 3 (September 2005): 235–248.

Tegmark, Max. *Our Mathematical Universe: My Quest for the Ultimate Nature of Reality*. New York: Random House, 2014.

———. "Our Mathematical Universe." Lecture given at the Royal Institution on January 30, 2014. Accessed March 18, 2019 at https://soundcloud.com/royal-institution/max-tegmark-our-mathematical.

———. "The Mathematical Universe." *Foundations of Physics* 38, no. 2 (February 2008): 101–150.

Thierry of Chartres. *Treatise on the Work of the Six Days*. Translated by Katharine Park. Accessed November 3, 2022 at https://www.academia.edu/31388090/Thierry_of_Chartres-Treatise_Six_Days-trans._Park.pdf.

Torrance, Thomas. *Christian Theology and Scientific Culture*. Eugene, OR: Wipf and Stock, 1980.

———. *The Ground and Grammar of Theology*. New York: T&T Clark, 1980.

Trigg, Roger. *Beyond Matter: Why Science Needs Metaphysics*. West Conshohocken, PA: Templeton Press, 2015.

Tyson, Neil deGrasse. *Astrophysics for People in a Hurry*. New York: W.W. Norton and Co., 2017.

van der Schoot, Albert. "Kepler's search for form and proportion." *Renaissance Studies* 15, no. 1 (March 2001): 59–78.

van Inwagen, Peter. "C.S. Lewis' Argument Against Naturalism," *The Journal of Inkling Studies* 1, no. 2 (October 2011): 25–40.

———. "Did God Create Shapes?" *Philosophia Christi* 17, no. 2 (2015): 285–290.

———. "God and Other Uncreated Things." In *Metaphysics and God: Essays in Honor of Eleonore Stump*, edited by Kevin Timpe. London: Routledge, 2009.

Voelkel, James. *Johannes Kepler and the New Astronomy*. New York: Oxford University Press, 1999.

Walsh, P. G. "Introduction," in Boethius, *Consolation of Philosophy*. New York: Oxford University Press, 2008.

Ward, Keith. *God, Chance & Necessity*. Oxford: Oneworld Publications, 2009.

Wegner, Daniel. *The Illusion of Conscious Will*. Cambridge, MA: MIT, 2002.

Weinberg, Steven. *Dreams of a Final Theory: The Scientist's Search for the Ultimate Laws of Nature*. New York: Vintage Books, 1994.

———. "Lecture on the Applicability of Mathematics," *Notices of the American Mathematical Society* 33.5 (Oct), quoted in Mark Steiner, *The Applicability of Mathematics as a Philosophical Problem*. Cambridge: Harvard University Press, 1998.

Welty, Greg. "Theistic Conceptual Realism: The Case for Interpreting Abstract Objects as Divine Ideas." D.Phil thesis, Oxford University, 2006.

Weyl, Hermann. *Mind and Nature*. Princeton: Princeton University Press, 2009.

Whitehead, Alfred North. *An Introduction to Mathematics*. Great Books of the Western World 56. Chicago: Encyclopaedia Britannica, Inc., 1990.

Wigner, Eugene and Andrew Szanton. *The Recollections of Eugene P. Wigner*. Cambridge, MA: Basic Books, 2003.

Wigner, Eugene. "The Unreasonable Effectiveness of Mathematics in the Natural Sciences." In *The World Treasury of Physics, Astronomy, and Mathematics*, edited by Timothy Ferris. Boston: Little, Brown & Co., 1991.

Wilbur, Richard. *Collected Poems: 1943–2004*. New York: Harvest Publishers, 2004.

Winston, David. *The Wisdom of Solomon: A New Translation with Introduction and Commentary*. New York: Doubleday, 1979.

Yandell, Keith. "God and Propositions," in *Beyond the Control of God?*, ed. Paul Gould. New York: Bloomsbury Academic, 2014.

GENERAL INDEX

absolute creationism, 182–184, 186–187

Adler, Mortimer, 13

Androcydes, 45

anima motrix, 86

Anscombe, Elizabeth, 253, 303–304

Antiochus, 41

anti-Platonist, 187, 192, 196

anti-realism, 181, 192–194, 198

Apollonius of Perga, 105, 215

    On Conic Sections, 105, 215

archetype, 5, 7–9, 14, 41, 46, 58–59, 69, 77, 79, 97–98, 100, 116, 119, 140, 149–152, 154–158, 160–161, 165–167, 172–174, 186, 197–199, 239, 281

Archytas of Tarentum, 16, 20–23, 34–35

Argument from Reason (AR), 241–242, 257–258, 263, 268–269, 274, 277

Aristotelian, 5, 33, 50, 69, 102, 126, 130, 156, 294

Aristotelian-Ptolemaic, xxi, 51, 81, 96, 121, 124, 133, 143

Aristotle, 16–17, 20–26, 34–35, 75, 70, 117, 119, 153, 294–297, 300

    Metaphysics, 23, 26, 297

    On the Heavens, 25

Aristoxenus, 21, 34

arithmetic, 22, 30, 45–46, 54, 56, 72, 78, 82, 90, 104, 111, 171, 297

    arithmetical, 31, 64, 296

aseity, 168, 170, 173, 187, 191, 194

See also *divine aseity*

aseity-sovereignty doctrine (AD), 168–169, 173

astronomer, 13, 18, 66, 82, 86, 92–93, 102, 112, 128, 133–134, 138, 141, 144, 146–147, 236, 288–289, 301

astronomy, xxiii, 3, 5, 7–8, 22, 24, 37–38, 45, 52, 54, 57, 65, 72, 78, 82–83, 87–88, 90–94, 98, 103–104, 111, 121–122, 124, 126, 130, 133, 135–138, 142, 145, 150, 155, 157, 161, 281, 299, 301

Athanasius, 59–62

Against the Heathen, 59

atheism, xvii–xviii

atheist/atheistic, xvii, 171, 208, 211, 216, 218, 233, 251, 272, 302

Augustine of Hippo, xxii–xxiii, 49, 61–67, 76–79, 82, 95, 114, 118–119, 166, 170–172, 198, 242, 300

Confessions, 63, 65

On Christian Doctrine, 63

On Music, 63

Sermons, 61–62

Balfour, Arthur, 249–250, 253, 255, 257

The Foundations of Belief, 249

Bartsch, Jakob, 288–290

Basil of Caesarea, 59, 61–62, 78

Big Bang theory, 209

black hole, 223, 303

Boethius, 43, 71–74, 77, 79, 242

Consolation of Philosophy, 73

De institutione arithmetica, 72

book of nature/*liber naturae*, 7, 59, 61–62, 98, 127, 133, 141–142, 206, 283

book of Scripture/*liber scripturae*, 62, 133

See also *Scripture*

General Index

Book of Wisdom/The Wisdom of Solomon, 42–43, 60, 66, 76, 82, 119

bootstrapping objection, 184

Brahe, Tycho, 93, 102–103, 113, 155, 286–287, 289

Calcidius, 71

Calvinism/Calvinist, 1, 3, 286

Caspar, Max, 2, 5, 13, 49, 57, 78, 92–95, 98, 108, 110, 114, 118, 121–122, 124, 134, 144, 150–151, 156–159, 281, 285, 289, 298, 300

Castelli, Benedetto, 127

Catholic/Catholicism, 1, 3, 78, 102, 286–288

celestial physics, xxiii, 5, 103, 122

  See also *physica coelestis*

CERN, 222

Christian, 5, 8, 36, 42, 49, 55–56, 59, 71, 75, 95, 100, 118, 141, 149, 154, 161, 165, 172, 198, 241, 278, 282, 286, 301

  faith, 152, 285

  humanist, 91

  Judeo-Christian, xviii–xix, xxiv, 57

  liberal arts tradition, 73

  monotheism, 66

  neo-Platonist, 72

  neo-Pythagoreanism, 78

  philosopher/philosophy, 72, 76, 165, 167

  Platonic, 78

  Platonism/Platonist, 34, 166, 194, 299

  pre-Christian, 15

  religion, 153

  theism/theist, 127, 139, 152, 165, 192

  theology, 119, 299

  West, xix, 27

Christianity, xviii, 56, 62–64, 67, 73, 88, 131, 139, 278, 299

Church Father, xxii, 56, 59, 62, 168, 198, 206, 299–300

Cicero, 3, 101

Clement of Alexandria, 55–57

   The Stromata, 55

Copernican, 4, 86, 92, 94–95, 130, 144, 281

   cosmology, 85

   cosmos, 84, 86

   model, 3, 124

   Revolution, 130

   theory, 122, 124

Copernicanism, 93, 102, 121, 159, 297–298

Copernicus, Nicolaus, xviii, xx, 8, 49, 82–84, 86, 88, 93, 121–122, 141, 143, 146, 157

   On the Revolutions of the Heavenly Spheres, 82, 93

Copan, Paul, 195

copy, 5, 7, 9, 100, 116, 119, 149–150, 157–158, 160–161, 166, 198–200, 225, 239, 281

cosmic comprehensibility/intelligibility, 5–7, 52, 160, 199–200, 238–239

   See also *mathematical: comprehensibility/intelligibility*

cosmic harmony, 5, 24, 54, 61, 111, 121, 149

   See also *harmony*

cosmogony, 18, 27, 38, 100, 294

cosmological constant, 270–272

cosmology, 7, 18, 25, 31, 34, 39, 42, 81, 95, 100, 114, 119–120, 122, 126, 129, 252

   Aristotelian-Ptolemaic, 51

   Copernican, 85

   geocentric, 75

   heliocentric, 143

   mathematical, 14, 46, 158

    Philolaic, 18

    philosophical, 293

    Platonic, 77, 100

    Ptolemaic, 50

    Pythagorean, 19

counter-earth, 18, 24–25, 31

Counter Reformation, 108

Coyne, George, 236–237

Crantor, 35

Craig, William Lane, 169, 175, 178, 186–187, 190–198

cube, 28, 96, 99

  See also *polyhedra*

Darwin, Charles, 217, 247–248

    Descent of Man, 248

    Origin of Species, 247–248

Darwinian, 226

    neo-Darwinian, 274–275

Davies, Paul, 276

Delian problem, 21

Demiurge, 27, 47, 101

Dennett, Daniel, 265–266, 269

Descartes, René, 129, 178, 242–244

    Meditations on First Philosophy, 242, 244

Diogenes Laertius, 17, 20–21

divine aseity, 168, 172, 188, 191, 194

  See also *aseity*

dodecahedron, 28, 96

  See also *polyhedra*

Dyad, 35, 47

Eddington, Arthur, 200, 208–212, 302–303
    Fundamental Theory, 210
    The Expanding Universe, 209
    The Mathematical Theory of Relativity, 208
    The Nature of the Physical World, 211–212
    Stellar Movements and the Structure of the Universe, 208

Einstein, Albert, 199, 205–208, 212, 216, 219–220, 222–223, 227, 230–231, 236, 270–272, 302

enkyklios paideia, 37

Euclid, 44, 68, 100, 118, 176, 183, 296
    Elements, 67, 72, 92, 95, 109, 153

Eudemus of Rhodes, 21–22

Eudorus, 34–36, 41

ex nihilo, 13, 27

Exodus, 114, 299

Forms, 26–27, 31, 25, 41, 44, 46, 69, 75, 79–80
    See also *Platonic Forms*

Formula of Concord, 94, 107–108, 285

Galilei, Galileo, xviii, xx, 81, 102, 107, 123–124, 126–128, 130–131, 133, 145–146, 215, 217, 234, 237, 289
    The Assayer, 127
    Discourses on Two New Sciences, 126
    Sidereus Nuncius (Sidereal Messenger), 123, 144

Genesis, xxii, 38, 41, 66, 77, 92, 120

geocentric/geocentrism, 3, 39, 50, 75, 92, 143, 145, 298

geometrical, 21, 28, 39, 50, 78, 96, 99–100, 104, 108, 116, 118, 120, 133, 140, 155, 175–176
    archetype, 155, 166–167
    concepts/ideas, 57, 165, 177, 189, 296

General Index

    figures, 69, 99, 128

    model, 5, 96, 111, 115, 122

    shapes, 165, 172, 189, 193

geometry, 21–22, 24, 37–39, 45, 54–56, 63, 69–70, 72, 78, 82, 90, 96, 99–101, 104, 109, 111, 119, 134, 137, 152, 167, 214, 216, 220, 297, 299

    Euclidean, 170, 231

    non-Euclidean, 216, 220

    Riemannian, 216, 220, 227

Gingerich, Owen, 83, 93, 103, 105, 112, 130, 133, 138

Goetz, Stewart, 256–257, 259, 268, 277

Gould, Paul, 168, 184–187, 193, 195, 198

Grand Duchess Christina, 127, 145

Grattan-Guinness, Ivor, 218–220

gravity, 104, 126, 205, 208

Graz, 4, 89, 94, 102

great ideas, 7, 13

Hafenreffer, Matthias, 3–4, 108, 300

Haldane, J. B. S., 251–252, 256, 303

    "Possible Worlds," 251

Hardy, G. H., 171–172

    A Mathematician's Apology, 171

harmonic law, 112, 114, 137–138, 141, 158

  See also *planetary laws: Third Law*

harmony, 5, 7, 17–18, 30, 42–43, 53–54, 59–61, 80, 82–84, 86–87, 108–110, 113–117, 137–139, 149, 151–152, 154, 158, 160, 167, 205, 207, 212, 294, 302

    celestial, 138

    cosmic, 5, 24, 54, 61, 111, 121, 149

    mathematical, 99, 121, 136, 159

See also *mathematical: harmonies/harmonization*

three-part/tripartite, 5, 115–116, 128, 149–150, 154, 157–160, 166, 199, 281–282

Heerbrand, Jacob, 91–92

heliocentric/heliocentrism, 52, 81–84, 86, 92–93 102, 121, 124, 135, 139, 143, 145–146, 155–157

Hellenistic, 36–37, 42

Heller, Michael, 236–237

Hicks, Andrew, 71–72, 74–75

Higgs, Peter, 222–223

Higgs field/Higgs field theory, 222–223, 226

Holy Spirit, 127, 146, 156

Huffman, Carl, 16, 20–22, 293–295

Iamblichus, 44, 58, 74

icosahedron, 28, 96, 99, 176

See also *polyhedra*

image, 5, 7, 9, 116, 149, 154, 157, 160–161, 189, 199, 241, 281

image of God/God's image, xx, xxiii, 5, 14, 40, 42, 58, 61, 67, 98, 116, 118, 131, 133, 149–150, 152–153, 156–157, 160, 166, 198–199, 239, 241, 243, 257, 278

image-bearers, 41, 76, 279

image of the divine, 144, 301

imago Dei, xxiii, 90, 153, 157, 274, 278

inertia, 104, 129

intelligibility, xix, 41, 236, 278

See also *cosmic comprehensibility/intelligibility* and *mathematical: comprehensibility/intelligibility*

Irenaeus, 168

John of Salisbury, 75–76, 82

Metalogicon, 168

General Index

Kant, Immanuel, 246–247, 254, 272
Kantian, 227, 246–247
Kepler, Friedrich, 107
Kepler, Heinrich, 1
Kepler, Johannes
    banishment from Graz, 102
    children, 103, 107
    district mathematician, 107–108
    early life and education, 1–3
    illness and death, 288–289
    Imperial Mathematician, 102–103
    marriage, 102, 285
    mathematical cosmology, 158
    natural philosophy/natural theology, 7–8, 13–14, 49, 54, 67, 75–76, 78, 89, 108, 116, 120, 138–139, 152, 154, 159, 282
    teacher, 4, 94
    university years, 3–4, 89–94
    works
        Astronomia Nova, 89, 102–105, 130, 134, 145, 156, 282
        Dioptrice, 126
        Eclogae Chronicae, 108
        Epitome Astronomiae Copernicanae (*Epitome of Copernican Astronomy*), 121–123, 126, 142, 156, 286
        Harmonice Mundi (*Harmony of the World*), xx, 68, 99, 107–110, 116, 121, 130, 135–136, 142, 282, 286, 301
        Mysterium Cosmographicum, 89, 94, 96–102, 105, 108, 111, 113, 115, 138, 141, 150–151, 155–156, 159, 167, 282
        Rudolphine Tables, 286–287
        Somnium, 81, 290
Kepler, Katharina, 1–2, 286
Kepler, Ludwig, 285, 290

Keplerian, xxiv, 7, 101, 109, 126, 154, 160, 174, 186, 199, 203, 239, 286, 290
    natural theology, 6–9, 160–161, 166–167, 181, 190, 200, 225, 239, 241, 274, 278, 281–282
    philosophy, 115
Kim, Jaegwon, 267–268
Kline, Morris, 215–216, 220–222

law of gravitation, 129, 216
Leftow, Brian, 168–169, 178, 185, 274
Leibniz, Gottfried, 205, 244–245
    Leibniz's Law, 265
Lennox, John, 278
Lewis, C. S., 71, 253–258, 261, 266, 272, 303–304
Lewisian, 257–258, 263, 268–269, 277
liberal arts, 3–4, 27, 43, 62–63, 73, 87, 90–91, 95, 297
liber naturae
  See *book of nature*
liber scripturae
  See *book of Scripture*
Linz, 107–108, 285–288
Locke, John, 244
    Essay Concerning Human Understanding, 244
logic, 218, 252, 258, 275
Logos, 1, 41–42, 47, 60, 80, 155, 169, 212
Lutheran/Lutheranism, 1–3, 89, 94, 108, 218, 236, 285

Maestlin, Michael, 3, 91–93, 98, 104, 150, 156, 297–298, 300
    Epitome Astronomia, 92
Martyr, Justin, 55
    Hortatory Address to the Greeks, 55

General Index

mathematical, 4, 21, 50, 52–53, 69, 72, 74–75, 77, 79, 90–91, 94–95, 97, 101, 105, 109–111, 115–116, 121, 126–127, 129–130, 134, 137, 151, 153–154, 157–158, 165–166, 171, 173, 195, 199, 202–203, 205, 207, 209, 216, 218, 221–222, 226–227, 230–234, 239, 246, 264–265, 271, 275, 277, 281-282, 289, 301

    archetype, 149, 151, 156, 158, 165–166, 172–174, 197, 199

    arts, 16, 22, 55, 90, 128

    astronomers/astronomy, 86–87, 133

    comprehensibility/intelligibility, 9, 18, 52, 199, 206, 235, 239

    concept/conception, 151, 158, 167, 172, 213–214, 219, 226, 230

    constants, 202

    cosmology, 14, 46, 158

    curriculum, 22, 72

    discipline, 45, 56, 65, 77–78

    discourse, 68, 175

    entities, 26, 166–167, 170, 229

    equation, 200, 209, 222, 271

    forms, 141, 195, 232

    functions, 18, 172

    harmonies/harmonization, 86–87, 111–112, 117, 149, 154, 204

    ideas, xxii, 69, 118, 154, 158, 165–167, 195, 220

    language, 127, 214, 217

    law, 135, 235, 273, 276

    model, 100, 120, 155, 159, 237

    notation, 226, 236

    object, 167, 172, 174, 178–179, 181, 188–189, 191–192, 194, 197–198, 233–234

    order/ordered/ordering/orderliness, 18, 29, 42, 75, 81, 90, 99, 115, 152–153, 160, 210, 212–213, 219, 231, 234, 281–282

    plan, 5, 14, 98, 116, 149, 165

    physics, 7, 129–130, 207

    proposition, 165, 183, 189–190, 196, 265

Pythagoreanism, 74

rationality, 127, 234

ratios, 22, 54

reasoning, 6–7, 128, 217, 220–221, 235, 241, 270

regularities, 30, 234

science, 38, 45–46, 72, 79, 90, 293, 297

structure, 172, 194, 220, 233

   of nature, 131, 160, 201, 203, 212, 219

   of the cosmos, 14, 27, 93

   of the material realm, 6

   of the physical world, 194

system, 172, 214, 216, 226, 239

theories/theory, 52, 111, 199, 218–219, 222, 227, 230–231, 270, 278

truths, 172, 188–189, 198, 218, 229, 231–232, 254, 264, 271, 273, 282

universe, 128, 159, 234

world, 230, 233

Mathematical Universe Hypothesis (MUH), 233–235

mathematician, 5, 15, 21–23, 43, 93, 102, 107–108, 128, 136, 171, 206, 210, 214–216, 218–219, 221, 229, 251–252, 278, 287, 296

mathematics, 3–7, 9, 14, 20–21, 23, 26, 45, 52–54, 56, 63–64, 68–69, 72, 74, 77, 79, 89–94, 98, 100, 105, 109, 120, 126–127, 135, 137, 154, 160, 166–167, 170–172, 183, 191, 194–195, 200, 205, 209–210, 212–223, 225–239, 246, 250–252, 261, 271, 273, 275–276, 278, 281–282, 297

mathematics-nature-mind resonance, 6, 8

Melanchthon, Philip, 89–92, 297

Menuge, Angus, 189, 273

Methuen, Charlotte, 78, 90–91, 297–298

Modified Theistic Activism (MTA), 184–185

Monad, 35, 47, 245

General Index

Moreland, J. P., 261, 269, 270, 273
Moses, 36, 77, 120
Müller, Barbara, 102
Müller, Philipp, 288–289
music, 22, 37–39, 45, 54–56, 63, 74, 78, 82, 109–110, 299
    of the spheres, 24, 31, 110
    theory, 46, 65
musical, 38, 116, 137, 151
    harmonies, 15, 54
    intervals, 31
    law, 65
    numbers, 53
    ratios, 115,
    scales, 22, 24, 111
musician, 60, 86

Nagel, Thomas, 272
natural philosopher, xxiii, 5–7, 50, 75, 77, 130, 141, 143, 149, 152, 154, 279, 281
natural philosophy, 1, 6–7, 9, 13–14, 52, 68, 75, 86, 90, 95, 104, 127, 130–131, 133, 139, 141, 146–147, 151, 154
natural theology, 6, 91, 133, 141, 143, 159, 161
naturalism, xvii–xviii, 199–200, 225, 227–228, 232, 235, 238, 241–242, 250, 253–261, 264–266, 268, 272–274, 277, 282–283, 304
naturalistic evolution/naturalistic evolutionary, 227, 249, 265, 274
neo-Darwinian, 274–275
neo-Platonism, 5, 36, 57–58, 62, 67, 157
neo-Platonist, 5, 22, 44, 59, 62, 72, 78, 92
neo-Pythagoreanism, 33, 58, 78
Newton, Sir Isaac, xviii, xx, 86, 104–105, 122–123, 128–131, 135–136, 205, 207, 216, 220, 237, 246, 289, 300

Mathematical Principles of Natural Philosophy (*Principia*), 128, 130
Newtonian, 105, 115, 126, 128–130, 159
Nicholas of Cusa, 49, 58, 78–81, 92, 95, 156
    On Learned Ignorance, 78
Nicomachean, 47
Nicomachus of Gerasa, 43–46, 74, 119
    Introduction to Arithmetic, 43–44, 67, 72
    Introduction to Geometry, 44
    Introduction to Harmonics, 44
    Life of Pythagoras, 44
    Theology of Arithmetic, 44, 47
Nobel Prize, 200, 205, 212, 223
nominalism, 192–197
non-theist, 171–172, 199, 225, 228, 282
nous, 58, 69

octahedron, 28, 96, 99
  See also *polyhedra*
Origen, 56, 299–300

paideia, 56
pantheism/pantheistic, 212, 302
Parrish, Stephen, 189–190
Penrose, Roger, 228–233
    Penrose's Triangle, 228, 232
Philo Judaeus (Philo of Alexandria), 34, 36–43, 55, 66, 119, 166
    On Mating with the Preliminary Studies, 37
    On the Creation, 38
    Who is the Heir of Divine Things, 39
Philolaus, 16–18, 20–21, 23, 34, 152, 294–296

General Index

Philolaic, 20, 295–296
    cosmology, 18
Philonic, 36, 46, 294
philosophy, xxii, 7, 14–16, 20–21, 23, 30, 33, 36–37, 42, 44, 50, 52, 55–56, 73, 90–91, 117, 127, 142, 211, 241, 282, 297, 299–300
    archetypal, 111
    Aristotelian, 5
    Christian, 165
    contemporary, 257, 266
    mechanistic, 81
    of nature, 119, 265
    pre-Socratic, 16
    Pythagorean, 17, 23, 120
    Pythagorean-Platonic, 23, 95, 128, 282
    "two books," 62, 127
physica coelestis, xxiii, 133, 281
Planck, Max, 199–206, 208, 212
    Planck's constant, 200
    Scientific Autobiography and Other Papers, 201
planetary laws, 104, 106, 121, 124, 126, 129, 141
    First Law, 105–106, 124, 129–130, 138, 216
    Second Law, 104, 106, 130, 138
    Third Law, 105, 111–115, 129, 130, 138, 141, 149, 298–299
        See also *harmonic law*
Plantinga, Alvin, 173, 188, 190, 198, 273, 275–276
Plato, 5, 8, 16–17, 20–23, 26–27, 30, 31, 33, 37, 49, 53–54, 57, 67, 71–72, 76–78, 95, 99–102, 117–118, 120–121, 128, 152–153, 167, 187, 215, 294, 296
    Meno, 118, 153
    Phaedo, 17, 28
    Seventh Letter, 20

The Republic, 20, 22

Timaeus, 21, 27–31, 34, 53–54, 71–72, 77, 92, 99–102, 117–118, 120

Platonic, 26–27, 33, 36, 39, 58, 68, 71, 78, 92, 120, 140, 152, 154, 165, 167, 173, 176, 184, 186–187, 191, 220, 229, 232

    cosmology, 77, 100–101

    doctrine, 67, 118, 153

    Forms, 35, 41, 66, 68, 76, 80

    polyhedra, 176–177, 183, 197

    realism, 232

    solids, 28–29, 96, 100, 108, 113, 118–119, 177, 191

        See also *polyhedra*

    thought, 282

    triad, 47

    World Soul, 72, 77, 80

Platonism, 23, 26, 33–34, 37, 44, 57, 67, 71, 120, 165–166, 173–174, 178, 182, 190, 195, 229, 231, 233

    Christian, 166

    Middle, 8, 33–35, 42, 49, 57

    neo-Platonism, 5, 36, 57–58, 62, 67, 157

    theistic, 173–175, 181, 187, 195

Platonist, 34, 55, 77, 79–80, 165, 167, 172, 175, 178, 183, 194–195

    anti-Platonist, 187, 192, 196

    Christian, 34, 194, 299

    Middle, 5, 36, 41, 43

    neo-Platonist, 5, 22, 44, 59, 62, 72, 78, 92

Plotinus, 57–59, 62, 67

    Enneads, 58, 67

Plotinian, 58

Plutarch, 34, 101, 152, 167

Polkinghorne, John, 238

polyhedra, 15, 96, 98–100, 118, 121, 137–138, 176
>   cube, 28, 96, 99
>   dodecahedron, 28, 96
>   icosahedron, 28, 96, 99, 176
>   octahedron, 28, 96, 99
>   tetrahedron, 28, 96, 99

polyhedral, 158
>   model, 97, 100–101, 111, 115
>   theory, 97-98, 115, 119, 136, 138, 296–297

polyhedral/harmonic, 135
>   model, 158-159
>   theory, 136, 155, 158

Porphyry, 58, 62, 67

Prague, 93, 102–103, 107–108, 205, 287, 289

pre-Copernican, 81

pre-Platonic, 15, 293

pre-Socratic, 8, 14–16, 294

Proclus Diadochus, 22, 44, 49, 58, 67–69, 74, 78, 92, 95, 109, 118, 120–121, 153–154
>   *A Commentary on the First Book of Euclid's Elements*, 67–68, 95, 109, 120, 153

Protestant, 3–4, 94, 213, 288

Psalm, 42, 59, 146

Psalmist, 56, 59, 83

psychē, 58, 69

Ptolemaic, 3, 5, 50–51, 83–84, 92, 123

Ptolemy, Claudius, 8, 22, 44, 49–50, 52–54, 69, 75, 83, 93, 110, 122
>   *Almagest*, 50, 122
>   *Harmonics*, 22, 52–54, 110–111

Pythagoras, 14–17, 20–22, 44, 49, 55, 73–75, 77–79, 92, 95, 100–102, 170, 293

Pythagorean, 13–16, 19, 21–24, 26–27, 29, 35, 44–46, 53, 55–56, 74–75, 84, 99, 110, 116, 118, 127–128, 233, 295–297
    cosmology, 31
    mysticism, 20, 136
    neo-Pythagorean, 33, 44, 57, 74–76, 90
    philosophy, 17, 120
    teaching, 55
    tetractys, 29, 121
    theorem, 170, 231
    thought, 15–16, 23, 26–27, 52, 296–297
    tradition, 13, 20, 22–23, 36
Pythagoreanism, 8, 16, 20, 23, 26, 33–34, 44, 71, 73–75, 282, 293–294
    neo-Pythagoreanism, 33, 58, 78
Pythagorean-Platonic, 5–6, 8, 13, 23, 27, 31, 36, 38, 43, 45, 49, 54–56, 60, 64–65, 71, 73–74, 76, 78, 88, 93, 95–96, 99, 114–116, 119–120, 123, 128, 136, 139, 141, 149, 152, 160, 165–166, 28–282
    philosophy, 95, 128
    tradition, 95, 115, 119–120, 123, 149, 152, 160, 165–166
Pythagorica, 44

quadrivium, 22, 72–73, 77, 81–82
quantum mechanics, 201, 206, 213, 252, 275
quantum theory, 200, 209, 211–212, 217, 219
Quine, W. V. O., 174
Quinean Indispensability Argument, 174, 195
    neo-Quinean, 174–175, 195

Reid, Thomas, 245–246
Reidian, 251
realism, 181, 195
relativity theory, 208, 213, 216, 275

General Index

Reppert, Victor, 257, 266
Reuttinger, Susanna, 285, 288, 290
Rheticus, Georg Joachim, 86, 151
    Narratio prima, 86, 151
rhetoric/rhetorician, 3, 37, 62–63, 94, 299
Rudolf II, 107–108
Runia, David, 40
Russell, Bertrand, 171–172
    Principles of Mathematics, 171

sacramental, 140, 155, 157
Scripture, 3, 6, 36, 42, 62, 64–65, 80, 127, 139, 145–146, 169, 299
Socrates, 16–17, 21, 28
Solovine, Maurice, 206, 208
Stobaeus, Johannes, 18
    Anthology, 18
Stoic, 34, 36
Schwarzschild, Karl, 223
Steiner, Mark, 220, 225–227
Stephenson, Bruce, 14, 99, 114–115, 135, 298

Tegmark, Max, 222, 232–234
telescope, 123, 126, 209
tetrahedron, 28, 96, 99
  See also *polyhedra*
theism, 6, 168, 188, 191, 204, 239, 244, 274, 278
    Christian, 127, 139, 152
theist, 171, 173, 178, 182, 194, 199–200, 204, 225, 248, 256, 273, 278
    Christian, 165, 192
        non-theist, 171–172, 199, 225, 228, 282
theistic, 5, 123, 130, 213, 218, 239, 276, 281

activism, 183–184
  anti-realist, 194
  Platonism/Platonist, 173, 174, 175, 178, 181, 187, 195
theistic conceptualism, 182, 187–191, 195, 197–198
Theory of Everything (ToE), 233
theory of general relativity, 206, 208, 272
Thierry of Chartres, 76–78
  Treatise on the Work of the Six Days, 77
Thirty Years' War, 286, 289, 301
three-part harmony/tripartite harmony, 5, 115–116, 128, 149–150, 154, 157–160, 166, 199, 281–282
Torrance, Thomas, xxi, 278–279
Trigg, Roger, 234, 259, 263, 270
Trinity, 80, 131, 156
trinitarian, 156, 186
Tübingen, 57, 89, 91–94, 282, 297–298
  University of, 2–3, 8, 89, 94, 285

van Inwagen, Peter, 175,–178, 184, 194, 258
Virgil, 3, 94, 113, 170 281
von Hohenburg, Herwart, 134, 141, 143, 150, 156–157

Weinberg, Steven, 216
Western Tradition, 5–7, 13–15, 170, 199, 282
Welty, Greg, 187, 190
Weyl, Hermann, 252
  Mind and Nature, 252
Wheeler, John Archibald, 223, 303
Whitehead, Alfred North, xx, 215–216
Wigner, Eugene, 200, 212–219, 221–222, 227, 234, 238
Wisdom of Solomon

See *Book of Wisdom*
witch trial, 108
World War I, 223
World War II, 213
Württemberg, 2–3, 107

Xenocrates, 35
    On Nature, 35

Yandell, Keith, 184–185, 188, 195–197

www.ingramcontent.com/pod-product-compliance
Lightning Source LLC
Chambersburg PA
CBHW050310120526
44592CB00014B/1857